TREE KANGAROO
CONSERVATION PROGRAM

ANNUAL REPORT 2014

EQUATOR PRIZE
CELEBRATING SUCCESS IN REDUCING POVERTY
THROUGH THE CONSERVATION
AND SUSTAINABLE USE OF BIODIVERSITY
UNITED NATIONS GENERAL ASSEMBLY
NEW YORK CITY
22 SEPTEMBER 2014

www.zoo.org/treekangaroo

Tree Kangaroo Conservation Program
Annual Report 2014
Seattle, Washington, USA • 2015

Contents

History

The Evolution of a Global Model for Conservation

The Tree Kangaroo Conservation Program (TKCP) is Woodland Park Zoo's signature international conservation program, focused on conserving the endangered Matschie's tree kangaroo (*Dendrolagus matschiei*) and the habitat in which it lives. From its beginnings in 1996 to determine the status of the endangered Matschie's tree kangaroo in the wild, TKCP has evolved from a species-specific conservation initiative into a holistic program supporting habitat protection for a wide range of threatened species, as well as initiatives to enhance local community livelihoods and access to government services.

Over the course of nearly two decades, the program has grown from its mountainous roots to embrace a broad landscape encompassing marine and coastal reef ecosystems, lowland and montane rain forests, alpine grasslands, and the agricultural areas and settlements belonging to more than 50 villages within the Yopno, Uruwa and Som (YUS) watershed

areas on the Huon Peninsula. Under PNG's customary land tenure system in which local people own and control over 90% of the land, long-term habitat protection relies on the commitment and participation of the local communities who depend on the forest's products and services. Together with local landowners and communities, TKCP partners with the PNG government, conservation biologists, social scientists, universities, research institutions, and other NGOs to build local capacity for the sustainable management of the YUS Conservation Area (YUS CA) and the surrounding landscape. Through these partnerships, local residents build a strong connection between their commitment to conservation in YUS and better opportunities for their families and communities.

With 18 years of effort and experience, TKCP has helped shape the concept of conservation for the country. The YUS CA sets a benchmark for which other PNG protected areas can aspire.

1996
TKCP begins its work in YUS

First land pledged for conservation by YUS Landowner Mambawe Manaono and clan

First tree kangaroo research in Dendawang, Sibidak and Pum Pum

2002
TKCP receives the AZA International Conservation Award

2004
First YUS Teacher's Scholarship students graduate and return to teach in YUS communities

First radio-collaring of tree kangaroos in Wasaunon

Wild tree kangaroo in YUS. *Photo by Bruce Beehler.*
YUS landscape. *Photo by Lisa Dabek, TKCP.*

About Us

Vision

The Tree Kangaroo Conservation Program envisions a sustainable, healthy, and resilient Huon Peninsula landscape which supports the area's unique biodiversity, human communities and culture.

Mission

The Tree Kangaroo Conservation Program fosters wildlife and habitat conservation and supports local community livelihoods in Papua New Guinea through global partnerships, land protection and scientific research.

2009

The YUS Conservation Area is formally gazetted by PNG's national government, protecting 78,000 hectares (87,000 acres) of wildlife habitat

The local landowner association is officially incorporated as the YUS Conservation Organization

YUS communities host an international celebration of the YUS CA gazettal

First National Geographic Society's Crittercam deployed on tree kangaroos

2011

The YUS Conservation Endowment is established at Woodland Park Zoo

First export of YUS Conservation Coffee to Caffe Vita in Seattle

Team of 12 YUS Conservation Area Rangers is established

YUS Ecological Monitoring Plan developed

2013

TKCP-PNG is registered as a local non-governmental organization in Papua New Guinea

YUS Landscape Plan approved by PNG government

2014

Community-led Land-use Planning is completed across all wards in YUS

TKCP receives the United Nations Equator Prize and its second AZA International Conservation Award

YUS Conservation Area Bylaws gazetted by PNG's national government

CP's efforts amplified with support of large grants from Conservation International *(2005-2014)* and the German Development Bank (KfW) *(2008-2014)*

Three Organizations Working Together for Sustainability

The Tree Kangaroo Conservation Program is the umbrella name for the partnership of WPZ's TKCP and TKCP-PNG.

1 TREE KANGAROO CONSERVATION PROGRAM – PAPUA NEW GUINEA (TKCP-PNG)

An independent non-governmental organization registered in PNG. Based in Lae, PNG. TKCP-PNG is the implementing partner of TKCP.

2 WOODLAND PARK ZOO'S TREE KANGAROO CONSERVATION PROGRAM (TKCP)

A program of Woodland Park Zoo's Field Conservation Department. Based in Seattle, Washington, USA.

3 YUS CONSERVATION ORGANIZATION (YUS CO)

A community-based organization representing the interests of local landowners and their communities. Based in YUS, PNG. YUS CO works in collaboration with TKCP.

Common Acronyms

AZA – Association of Zoos & Aquariums (USA)

CA – Conservation Area

CAMC – Conservation Area Management Committee

CBO – Community-based Organization

CEPA – Conservation Environment Protection Authority (PNG government)

DEC – Department of Environment and Conservation (now called CEPA)

IUCN – International Union for the Conservation of Nature

LLG – Local Level Government

LUP – Land-use Planning

NGO – Non-governmental Organization

PNG – Papua New Guinea

TKCP – Tree Kangaroo Conservation Program

TKCP-PNG – Tree Kangaroo Conservation Program – Papua New Guinea (local NGO)

UNDP GEF-SGP – United Nations Development Programme Global Environment Facility Small Grant Programme

WPZ – Woodland Park Zoo

YUS – Yopno-Uruwa-Som

YUS CO – YUS Conservation Organization (landowner association and community-based organization)

Biodiversity of YUS

1. **Dwarf Cassowary** – Muruk
   (Casuarius bennetti)
   IUCN Status: Near Threatened

2. **New Guinea Harpy-Eagle** – Bikpela Tarangau
   (Harpyopsis novaguineae)
   IUCN Status: Vulnerable

3. **New Guinea Vulturine Parrot** – Koki Red na Blak
   (Psittrichas fulgidus)
   IUCN Status: Vulnerable

4. **Emperor Bird of Paradise** – Kumul Waitpela Gras
   (Paradisaea guilielmi)
   IUCN Status: Near Threatened

5. **Huon Astrapia Bird of Paradise** – Blakpela Kumul Longpela Tel
   (Astrapia rothschildi)
   IUCN Status: Least Concern

6. **Wahnes's Parotia Bird of Paradise** – Blakpela Kumul Save Dens Long Graun
   (Parotia wahnesi)
   IUCN Status: Vulnerable

7. **Matschie's or Huon Tree Kangaroo** – Redpela Kapul
   (Dendrolagus matschiei)
   IUCN Status: Endangered

8. **New Guinea Pademelon** – Bikpela Sikau
   (Thylogale browni)
   IUCN Status: Vulnerable

9. **Eastern Long-beaked Echidna** – Nilnil Kapul Bilong Graun
   (Zaglossus bartoni)
   IUCN Status: Critically Endangered

10. **Spectacled Fruit Bat** – Bikpela Blak Bokis
    (Pteropus conspicillatus)
    IUCN Status: Least Concern

11. **Monitor Lizard** – Kundu Palai
    (Varanu sp.)
    IUCN Status: Least Concern

12. **Jewel Weevil**
    (Eupholis sp.)
    IUCN Status: Unknown

13. **Ulysses Swallowtail or Blue Emperor Butterfly**
    (Papilio Ulysses)
    IUCN Status: Unknown

14. **Long-fingered Triok** – Liklik Kapul Igat Longpela Pinga
    (Dactilopsila palpator)
    IUCN Status: Least Concern

Illustrations by: Stephen D. Nash, Conservation International from Wildlife of the YUS Conservation Area Pocket Identification Guide

Tree Kangaroo
Conservation Program Team

Seattle, Washington

Dr. Lisa Dabek	Program Director
Mr. Trevor Holbrook	Program Coordinator

Lae, Papua New Guinea

Ms. Mikal Nolan	Program Manager
Mr. Karau Kuna	Conservation Strategies Manager
Mr. Benjamin Sipa	Community Services and Livelihoods Manager
Mr. Danny Samandingke	Leadership Training and Outreach Senior Coordinator
Mr. Daniel Okena	Research and Monitoring Coordinator
Ms. Melanie Palili	Healthy Community Coordinator
Ms. Laris Bartsaka	Administrative Coordinator
Mr. Namo Yaoro	Conservation Officer
Mr. Dono Ogate	Conservation Officer
Mr. Steven Fononge	Conservation Officer
Mr. Chris Max	Conservation Officer
Mr. Victor Eki	Mapping Officer
Mr. Matthew Tombe	Mapping Officer

Front row, from left: Mikal Nolan, Namo Yaoro, Danny Samandingke; *Middle row, from left:* Laris Bartsaka, Melanie Palili, Victor Eki, Chris Max; *Back row, from left:* Daniel Okena, Benjamin Sipa, Dono Ogate, Matthew Tombe, Karau Kuna, Nelson Teut; *Not pictured:* Steven Fononge

 Top: TKCP Director Lisa Dabek and TKCP Coordinator Trevor Holbrook. *Photo by WPZ.*
Bottom: TKCP-PNG team in Lae. *Photo by TKCP.*
Right, top: YUS CA Ranger team. *Photo by TKCP.*
Right, bottom: TKCP Director Lisa Dabek with Mr. Mambawe Manaono. *Photo by TKCP.*

YUS Conservation Area Rangers

Mr. Soya Werawe	Kumbul village
Mr. Nelson Teut	Gua/Teptep village
Mr. Tamina Pendeng	Mek/Nolum village
Mr. James Jio	Towet village
Mr. Moses Nasing	Yawan village
Mr. Geno Yuwoc	Worin village
Mr. Robson Soseng	Gomdan/Sapmanga village
Mr. Muks Uwate	Gogiok village
Mr. Danny Wande	Kalaset village
Mr. Mike Barup	Ronji/Koripon village
Mr. Dogem Mirande	Bonkiman/Wadabong village
Mr. Sulu Mondo	Bonea village
Mr. Hesiya Manok	Singorokai village

Elected by local landowners to conduct monthly patrols through steep, treacherous terrain and harsh weather conditions, the YUS Conservation Rangers embody the steadfast commitment of local communities to protect their environment. TKCP is honored to assist Papua New Guinea's first Conservation Rangers in ecological monitoring, educating their communities on the rules and boundaries of the YUS Conservation Area, and enforcing the YUS CA bylaws in coordination with local government officers and magistrates. In addition to their traditional knowledge of the environment, the Rangers are quickly becoming local leaders in species identification and tracking, research assistance and conservation techniques.

TKCP and the YUS partners celebrated many exciting accomplishments during 2014, due to the expertise and dedication of the team in Lae and YUS, Papua New Guinea and Seattle, Washington. TKCP also welcomes several new team members:

- Trevor Holbrook joined as the Program Coordinator, focusing on program development, grant management and outreach in the U.S.

- Daniel Okena, the new Research and Monitoring Coordinator, is leading the ecological monitoring, providing direct support to the YUS Rangers, and coordinating scientific research efforts in YUS.

- Melanie Palili is leading TKCP's health services and education initiative as the Healthy Community Coordinator.

- Chris Max brings his knowledge of YUS coastal areas to engage with communities and YUS CA Rangers as TKCP's fourth Conservation Officer. Chris is a YUS landowner from Ronji village.

- Hesiya Manok joins as TKCP's first YUS CA Marine Ranger. Hesiya is a YUS Landowner from Singorokai village.

SPECIAL THANKS

We also recognize the contribution of TKCP Pioneer and Advisor, Mr. Mambawe Manaono – one of TKCP's first partners in YUS.

Visitors to the YUS Conservation Area in 2014

TKCP and YUS Communities were happy to welcome several of our colleagues and partners to the YUS CA, including:

Carol Esson, DVM

Cairns, Australia—Field Veterinarian and PhD candidate in animal health and conservation. Carol works as a field veterinarian with TKCP's tree kangaroo research. *See Strategy Two, page 20 for further details.*

Volunteer Physician Team

Dr. Robert Liddell (Center for Diagnostic Imaging and WPZ Board Member), Dr. Marti Liddell (Seattle Polyclinic), Drs. Blair Brooks, Nancy Philips, and Travis Austin (Dartmouth-Hitchcock Hospital), and Sister Baleb Wahazoka (Lae) visited Sapmanga village to conduct a health workshop with Community Health Workers and Village Birth Attendants from across YUS. The physicians also spent several days seeing patients in several YUS communities, providing diagnoses and basic treatment.
See Strategy Four, page 28 for further details.

Daniel Shewmaker

Coffee Buyer for Seattle, Washington's Caffe Vita—Daniel visited farmers in the Yopno zone to participate in a coffee training led by TKCP Livelihoods Manager Ben Sipa and the Coffee Industry Corporation's John Kabuba and Daniel Kilagi. Daniel explored Yopno's coffee gardens and collected a coffee sample for further evaluation in Seattle. *See Strategy Four, page 26 for further details.*

Chris Banks

International Conservation Partnerships Manager, Zoos Victoria – To support the development of a new partnership between Zoos Victoria and TKCP, Chris visited the Uruwa zone to meet with coffee farmers and gain firsthand knowledge of the impact of the program's livelihood initiatives.

Dr. Valerie Guerin

James Cook University—A postdoctoral research associate with JCU's Language and Culture Research Centre, Valerie is undertaking a multi-year study entitled "Language documentation and description of Tiatuk, the language of Gogiok."

Charlotte Jennings

University of California Berkeley—PhD candidate with UCB's Museum of Vertebrate Zoology and Department of Integrative Biology. Charlotte is studying the comparative thermal physiology and evolution of Australopapuan skinks.

Dr. Richard Donaghey

A biodiversity consultant based in Australia, conducting independent research on birds including mid- to upper-montane robins (Petroicidae) endemic to New Guinea. Richard was accompanied by research volunteer David Bryden, volunteer nature sound recordist Tony Baylis and volunteer ornithologist Donna Belder.

Gabriel Porolak, James Cook University

A PhD candidate conducting conservation research on the evaluation of Hunting Sustainability and the Potential Role of Protected Areas in Papua New Guinea. Gabriel has made several visits to YUS for his PhD research since 2012. Previously, he was awarded a Master's degree from JCU for the tree kangaroo home range and habitat study. (http://journals.plos.org/plosone/article?id=10.1371/journal.pone.0091870) Regretfully, his study was interrupted by illness. TKCP wishes him a full and speedy recovery.

In Memory of Curator Larry Collins, TKCP Inspiration

The tree kangaroo community will miss Larry Collins, retired curator at the Smithsonian Institution/National Zoo's Center for Research and Conservation (CRC) in the USA. Larry's interest in the Matschie's tree kangaroo led to years of behavioral and reproductive research, and he was instrumental in establishing the AZA Tree Kangaroo Species Survival Plan. His work, and his mentorship of others interested in expanding upon his work, resulted in improved husbandry and reproductive success in tree kangaroos. All that has been accomplished in the zoo profession with tree kangaroos, especially for Matschie's tree kangaroo, can be connected to Larry Collins.

Larry Collins, along with Judie Steenberg, was a mentor to Lisa Dabek from the early days of her tree kangaroo research. Lisa spent six months doing her dissertation research at the CRC in Front Royal, Virginia. Under the guidance of curator Larry Collins, she studied the reproductive biology of the captive breeding group of 20 Matschie's tree kangaroos. It was during this time that Larry put the idea of working in PNG into Lisa's mind. He said, "You will someday go to PNG." Lo and behold, Larry was right. Thank you, Larry, for inspiring the creation of the Tree Kangaroo Conservation Program.

Larry Collins is now memorialized in the YUS rain forests of Papua New Guinea through TKCP's research program. One of our latest tree kangaroos radio-collared in 2014 is named "Larry." He is a tough older tree kangaroo that ventures into "ples no gut" – the most difficult part of the jungle. Larry always was a champion of forging ahead, meeting challenges, and creating successful breeding programs for little-known species. He was proud that his work on tree kangaroo conservation extended into the wilds of the Huon Peninsula in Papua New Guinea. TKCP and the tree kangaroos thank you.

We will miss him dearly, yet his legacy will live on in all the work we do to conserve tree kangaroos and support the people of YUS. Larry will fondly be remembered by his favorite phrase, "Tree kangaroos are above it all."

From left to right:
Chris Banks of Zoos Victoria, left, with TKCP's Benjamin Sipa in Uruwa. *Photo by TKCP.*
Gabriel Porolak, right, on a field research visit. *Photo by Bruce Beehler.*
Tree kangaroo "Larry" collared during Lisa Dabek's 2014 research visit. *Photo by Lisa Dabek, TKCP.*

Awards and Recognition

2014 was an incredible year for the Tree Kangaroo Conservation Program as we celebrated milestones and received prestigious awards which represent the dedication and cumulative efforts of the YUS community over the course of nearly two decades.

United Nations Equator Prize

The United Nations Development Programme's Equator Initiative seeks to recognize and advance local sustainable development solutions for people, nature and resilient communities. Selected from among 1,200 applicants from more than 120 countries, TKCP was one of 25 programs worldwide to receive the 2014 Equator Prize in recognition of our holistic, community-owned approach to habitat conservation. Noting the program's success in facilitating cooperation and mobilizing action among the diverse indigenous clans of YUS to establish a community-owned Conservation Area, TKCP also received a Special Recognition Award for Sustainable Forest Management.

In September, TKCP's Lisa Dabek and Karau Kuna traveled to New York, USA to attend the Equator Prize Award Ceremony. Held at the famed Lincoln Center during the UN Secretary General's Climate Summit and World Conference on Indigenous Peoples, the event drew many leading conservationists, diplomats and prominent advocates. The ceremony included thought-inspiring speeches and calls to action by globally-recognized leaders including Dr. Jane Goodall, Al Gore and Jeffrey Sachs. With the award, TKCP joins the Equator Initiative's World Indigenous Network (WIN), along with other innovative community-led conservation programs. In New York, Dabek and Kuna had the opportunity to meet with the network of colleagues to share ideas and to learn from the vast wealth of grassroots-level experience represented.

Association of Zoos & Aquariums International Conservation Award

Conservation is the utmost priority for the Association of Zoos & Aquariums (AZA). Each year, the association presents its International Conservation Award to an AZA-accredited organization demonstrating exceptional efforts toward habitat preservation, species restoration and support of biodiversity in the wild. In 2014, Woodland Park Zoo and its AZA-affiliated partners received the award with top honors in recognition of TKCP's achievements in YUS. WPZ and TKCP are proud to be considered conservation leaders among AZA institutions. WPZ shares this award with its AZA partners: Brevard Zoo, Cleveland Metroparks Zoo, Columbus Zoo and Aquarium, Gladys Porter Zoo, Milwaukee County Zoological Gardens, Minnesota Zoological Garden, Oregon Zoo, Riverbanks Zoo & Garden, Roger Williams Park Zoo, Saint Louis Zoo, San Diego Zoo, Santa Fe College Teaching Zoo, Sedgwick County Zoo, Smithsonian National Zoological Park, and Zoo New England.

Partnership Affirmation – Woodland Park Zoo and TKCP-PNG

In recognition of the milestone establishment of TKCP's local non-governmental organization in Papua New Guinea, Woodland Park Zoo's Board Chair, Nancy Pellegrino, and President and CEO, Dr. Deborah Jensen, presented TKCP-PNG with a congratulatory plaque and letter affirming the continued partnership between the two organizations. The letter commended TKCP-PNG for its award-winning program, and recognized the incredible value of YUS for the protection of biodiversity and as a model of community-based protection in Papua New Guinea and abroad. Dr. Robert Liddell, member of the WPZ Board of Directors, presented the plaque to TKCP-PNG during his visit to PNG in October.

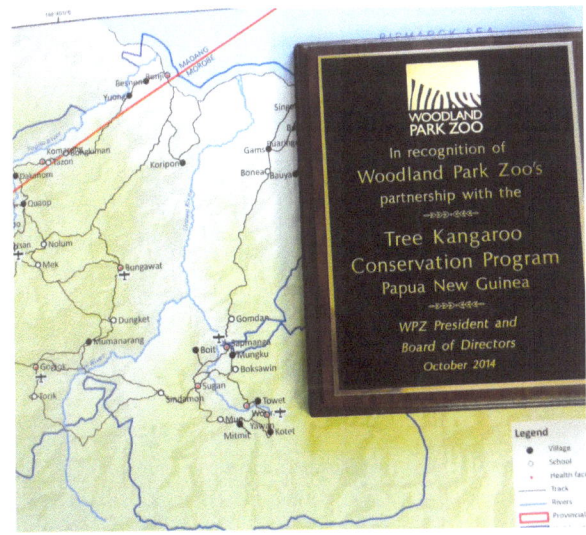

Clockwise from bottom left:
TKCP Conservation Strategies Manager Karau Kuna accepting the United Nations Equator Prize in New York City. *Photo by UNDP.*
United Nations Equator Prize trophy. *Photo by Mikal Nolan, TKCP.*
AZA International Conservation Award plaque. *Photo by WPZ.*
WPZ Partnership Affirmation plaque and map of YUS. *Photo by Dr. Rob Liddell.*

Progress Across the Living Landscape of YUS

The impact of the YUS Conservation Area gained substantial momentum during 2014, strengthened by the support of the PNG National Government and bolstered by the pride and commitment of the YUS people. Developed together with the communities, the formal gazettal of the YUS CA bylaws establish clear rules and regulations for the enforcement of the area's protected status. The development of the final ward-level land-use plans also ensured ecosystem connectivity across the YUS landscape. Completion of the land-use plans will enhance community input into the YUS Landscape Plan 2016-2020.

The YUS Landscape Plan guides the management and direction of the YUS Conservation Area and TKCP's efforts throughout YUS. Recognizing the success of the community-led land-use planning initiative in YUS, PNG's Department of Environment and Conservation has utilized TKCP's model for inclusion in the country's new Protected Areas Policy. Drawing upon the land-use planning model applied in YUS, the government aims to strengthen other protected areas throughout the country using TKCP's model of community engagement and local ownership.

Women walking to Sapmanga village. *Photo by Lisa Dabek, TKCP.*

THE FIVE STRATEGIES FOR ACHIEVING TKCP'S MISSION

ONE: Managing the YUS Conservation Area

TKCP works alongside the Conservation Area Management Committee (CAMC) in managing the YUS CA. Conservation Area Rangers conduct patrols and ecological monitoring, and check for violations within the YUS CA. Core protected areas are mapped by Mapping Officers. Local communities are informed and engaged in the protection of the YUS Conservation Area.

TWO: Applying Our Research

A broad range of research topics are investigated by TKCP, partners, and outside researchers to inform the direction of conservation efforts across the YUS landscape. In addition to TKCP's ongoing studies of tree kangaroo home range and behavior, other research studies examine ecological and social questions to assess needs, define responses, evaluate impact, and contribute to the global scientific knowledge of wildlife species, ecosystems and anthropology of YUS.

THREE: Planning for a Sustainable Future

Protecting the biodiversity and habitat of YUS requires coordinated commitment and action across the entire landscape, both in and around the YUS Conservation Area. To sustain the needs of local communities, the natural resources and services provided by the environment beyond the protected area must be maintained for the benefit of current and future generations. Managing the responsible use of the forest products, wildlife, and water in these areas will ensure the YUS communities' continued commitment to protecting the YUS Conservation Area.

Defined together with communities and landowners, the five strategies of the YUS Landscape Plan guide the conservation and socioeconomic development efforts of TKCP and our partners to ensure the sustainable health and prosperity for the living YUS landscape, biodiversity, people and culture.

FOUR: Supporting the Communities

The people of YUS rely on the natural environment for their day-to-day needs. TKCP works with communities to address their need for sustainable livelihoods as well as access to health, education and skills training. TKCP builds partnerships to provide YUS communities with alternative opportunities, which build local resilience and reduce the threat of short-term financial gain through large-scale resource extraction.

FIVE: Operating the Tree Kangaroo Conservation Program

The YUS conservation initiative attracts substantial worldwide attention and support. Managing such a program requires robust planning, administrative capacity and highly competent staff. TKCP remains committed to building the resources and staff capabilities to maintain long-term support for the landscape and people of YUS.

Managing the YUS Conservation Area

Governing the YUS Conservation Area

Following the gazettal of the YUS Conservation Area in 2009 and the YUS Landscape Plan in 2013, the protection of biodiversity in YUS has been further strengthened in 2014 with the national gazettal of the YUS Conservation Area bylaws. Developed by the landowners, the bylaws formally establish the rules within the protected area and define their associated penalties for application by the village court system. TKCP's Danny Samandingke provided a series of informational workshops throughout YUS to familiarize local leaders and law enforcement with the new bylaws and how to handle offenders. With the standardized bylaws and consistent penalties, communities recognize that the YUS CA is supported by government at all levels, and landowners are further assured that their conservation commitments will be respected and enforced by authorities.

> When I heard of the YUS CA bylaw induction workshop to be conducted in my ward area, I was glad to participate... I am fortunate to have been involved in the induction workshop, which has helped me to better understand the bylaws of the YUS CA. With a better understanding of the bylaws, I am better able to mediate bylaw violation cases, and I am in support of working with the YUS CA stakeholders.
>
> *Mr. Forenu Dakop, Village Court Magistrate, Ward 7, Nayudo LLG, Rai Coast District, Madang Province*

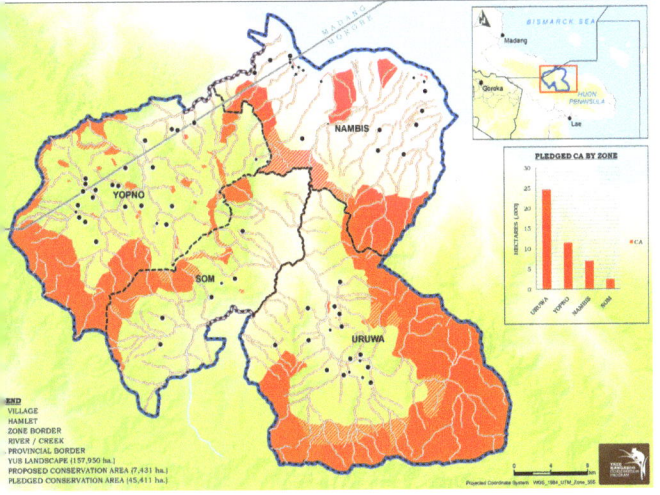

Front row, from left: Mr. Epemu Kiwenu, Mr. Petrus Yasing; Back row, from left: Mr. Timmy Sowang, Mr. Robin Kiki, Ms. Madeline Lahari, Ms. Laris Bartsaka, Ms. Annie Ogate, Mr. Botty Kwisal, Mr. Benside Thomas

The YUS Conservation Area Management Committee represents all stakeholders in the decision making for long-term management of the protected area. The committee's members include:

- YUS landowners
- Kabwum District Administrator
- Morobe Provincial Government representative
- National Department of Environment & Conservation representative
- President of the Yus Local Level Government
- President of the Wasu Local Level Government
- Program Manager of Tree Kangaroo Conservation Program

MAPPING AND CONSERVATION AREA AWARENESS

Together with Conservation Strategies Manager Karau Kuna and TKCP's ward-level land-use planning efforts, Mapping Officers Victor Eki and Matthew Tombe continued to map the YUS Conservation Area during 2014 in order to confirm boundaries, and to lead community awareness of the value of land pledges to further enhance connectivity throughout the YUS CA. 3,190 hectares of pledged land were mapped during 2014, including 46 hectares of coastal reef. In addition to mapping the YUS Conservation Area, TKCP team is mapping village land boundaries throughout YUS to prevent future disagreements and disputes.

In total, 45,411 hectares (112,213 acres) of YUS Conservation Area have been mapped, including 1,500 hectares of designated reforestation sites and 1,107 kilometers of 30-meter riparian corridors along the rivers and tributaries of YUS. Landowners continue to pledge additional land every year as a result of TKCP's successful community awareness and engagement.

YUS CONSERVATION AREA MANAGEMENT COMMITTEE

Entering its fifth year, the YUS Conservation Area Management Committee (CAMC) continues to play an integral role in encouraging communication among the various stakeholders of the YUS Conservation Area, from the community level to the provincial and national government level. The committee meets semi-annually to discuss key management issues and provide strategic direction for the YUS Conservation Area.

In April, the committee comprising of community and government representatives met in Lae to discuss a range of issues, including the upcoming approval of PNG's newly-revised Protected Areas Policy, the ongoing transition of the Department of Environment and Conservation (DEC) to the Conservation Environment Protection Authority (CEPA), and the future implications for the YUS Conservation Area. The second meeting was hosted by Isan village in YUS, and corresponded with a high-level visit from the Governor of Morobe Province and the Member of Parliament representing Kabwum District. The CAMC members met with the officials around the fire, sharing the successes and challenges facing the YUS Conservation Area. The opportunity to communicate directly with senior government representatives was thrilling for the committee members and Isan villagers, and demonstrated the ability of the CAMC to elevate the concerns and issues affecting YUS to the government representatives.

Clockwise from bottom left:
YUS Conservation Area bylaw induction ceremony. *Photo by TKCP.*
TKCP Mapping Officer Victor Eki. *Photo by TKCP.*
Mapped areas of the YUS Conservation Area as of 2014. *Map by Karau Kuna.*
YUS CA Management Committee. *Photo by Mikal Nolan, TKCP.*

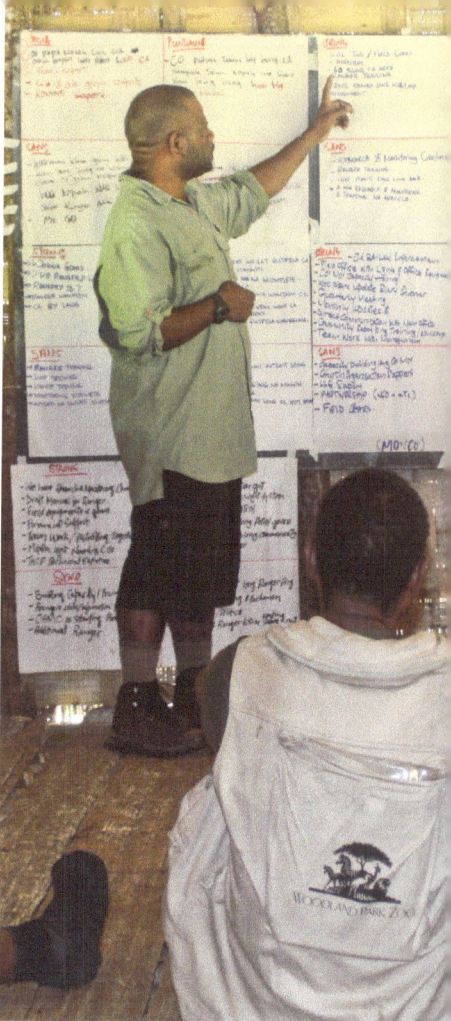

YUS Conservation Area Rangers

Established in 2012, the YUS Conservation Area Ranger team is responsible for patrolling and monitoring wildlife inside and outside of the YUS Conservation Area. The Rangers are local landowners, nominated by their own communities to conduct patrols and evaluate the effectiveness of conservation efforts in YUS. With the addition of a Marine Ranger in 2014, the team has expanded to 13 Rangers. Each month during 2014, the Rangers spent seven days patrolling segments of the YUS Conservation Area near their villages to record the presence or absence of several community-identified priority animal species key to achieving conservation goals, and to check for potential violations of the mutually-agreed bylaws governing the Conservation Area. Preliminary observations during the Ranger patrols throughout the year indicate increased sightings of previously-scarce wildlife, suggesting that the protected areas are having a positive impact for the biodiversity of YUS.

YUS CA RANGER FUNCTIONS

1. Conduct monthly patrols and gather data using the YUS Record Book

2. Serve as community liaisons to convey findings and advocate for conservation among local YUS communities

3. Participate in the YUS Ecological Monitoring Program

4. Provide field assistance for research conducted in YUS CA

5. Check for potential violations of bylaws in the YUS CA

My work area stretches along the reefs of Bugara, Singorokai and Ronji villages, including the protected areas of Bitex, Bukamu and the Utu reef area. What I saw with Mr. Job Opu during my monitoring patrol of the conservation area was unbelievable—marine life that we have never seen before! I am very excited to report what I encountered during my sea and reef patrols, which shows the success and impact of our conservation efforts. I am proud to be a YUS Ranger, and to share with the TKCP team and the coastal communities on the progress in the marine conservation area.

Mr. Hesiya Manok, YUS CA Marine Ranger
from Singorokai village, Nambis zone

MARINE RANGER MONITORING

The YUS Conservation Area ranges from coral reefs to 4,000 meter peaks. Thanks to funding received through the United Nations Development Program's GEF Small Grants Programme, TKCP contracted a marine ecologist, Mr. Job Opu, to enhance our understanding of the marine ecosystems and Ridge-to-Reef protection throughout YUS. Mr. Opu conducted a marine biodiversity survey and invited nearly 100 men and women from the coastal villages of Ronji, Singorokai, and Bugara to understand the marine environment and how it is used, and to identify environmental threats and the local means to address them. As a result of the workshops and TKCP's consultations with coastal villages, the communities have already committed to ending destructive fishing techniques, including the use of dynamite and mosquito nets. With training from Mr. Opu, TKCP's new YUS Marine Ranger is helping to protect the reefs and marine biodiversity of YUS, including nesting sites for the endangered leatherback turtle sighted during his first patrol.

From left to right:
YUS Conservation Area Rangers on patrol. *Photo by TKCP.*
TKCP's Benjamin Sipa (standing) facilitating YUS CA Ranger program evaluation. *Photo by Lisa Dabek, TKCP.*
YUS CA Marine Ranger Hesiya Manok, right, surveying coastal reefs in Nambis zone with marine specialist Job Opu. *Photo by Daniel Okena, TKCP.*
YUS CA sign alerting community members of 'tambu' reef area in Nambis zone. *Photo by TKCP.*

THE LIFE OF A YUS CONSERVATION AREA RANGER

A COMPILATION OF THE LIVES AND EXPERIENCES SHARED BY TKCP'S 13 YUS RANGERS

By Daniel Okena

Steep mountainous terrains, thick dense forest, and narrow walking tracks shifting due to landslides and floods. The environment I live in is extremely harsh and unforgiving, and may seem uninhabitable. But our people have been living here for many generations.

Here, we build a hut, plant a garden, and hunt for our food. We collect our water, firewood, building materials and food from the surrounding land. We teach our children how to do the same. A great thing about living in this environment – everything is free! We pay no rent or water bills, we do not have to buy our land, and our food and building materials are simply free. All we need to do is manage these resources sustainably.

The concept of conservation is not new to us. Traditionally, our village chiefs and elders were entrusted to be the stewards of the land; a great privilege and honor. Some of the land outside of our villages is declared off-limits, or 'Tambu' – these areas are believed to be controlled by powerful unseen beings. Trespassers and their families will be cursed with bad fortune or death. These Tambu areas then serve as reserves, regenerating and repopulating the entire forest.

It is a great honor to continue the legacy of stewarding my environment and its resources. Although the challenge is greater than it once was, TKCP has trained me to monitor changes in my unique forest, with its incredible terrains flowing with breathtaking creeks and rivers. As a YUS Conservation Area Ranger, I spend one week every month patrolling my environment. Equipped with a Global Positioning System (GPS) device, a pen and datasheet, safety boots, camping gear, and rations, I look for the presence of some of our forest's different animals. I also observe and record any signs of illegal activity within the YUS Conservation Area, and report violators to the landowner to take the matter before the local courts.

Patrolling in the CA is no easy task. Some areas are very far from the villages, with steep hills and mountains, deep valleys, and no walking tracks. But, as YUS Conservation Area Rangers,

we roam with the strong conviction that our efforts are for the long-term benefit of our people. We walk for miles through the wind, rain, and cold, and under the relentless heat of the sun to fulfill our honorable duties.

Apart from our Ranger work, we also have a family to look after. When we are not on patrol, we spend time with our families, tending our gardens and hunting in the hunting zone areas so that our families can enjoy life in this part of the world. Our patrol allowance from TKCP helps us to pay school fees for our children and support our extended family members.

It is great to be a YUS CA Ranger, but it is not without its challenges. The area is quite big, and it can be difficult to cover our entire patrol routes each month. But with the help from community members across YUS, we can all contribute by respecting the protected areas and exposing any violations. We must urge our neighbors to adhere so that our children and our children's children will have the privilege to enjoy the magnificent environment that we now enjoy.

From left to right:
YUS CA Ranger James Jio in the village with Karau Kuna. *Photo by Lisa Dabek, TKCP.* Hiking trail in YUS. *Photo by Carol Esson.*

Future Plans

- Equip Rangers and Conservation Officers with training, equipment, and appropriate technology to support YUS Conservation Area monitoring data collection and analysis

- Increase community awareness of YUS CA boundaries using informational materials and signposting

- Enhance the effectiveness of the YUS Conservation Area Management Committee

Applying Our Research

Tree Kangaroo Ecological Research

Research remains a central component of TKCP's work in YUS. Having studied wild Matschie's tree kangaroos in YUS since 1996, we continually seek to gain a greater understanding of the species we are trying to protect, and assess the effectiveness of our conservation efforts.

The Tree Kangaroo Conservation Program continues to study the dietary needs, behavior, and habitat use of Matschie's tree kangaroos at various elevations throughout their distributional range (1,000 to 3,500 meters). Building on our previous home range research conducted at 3,000 meters, TKCP has established a second research site at a lower elevation of 1,500 meters. The research will inform decisions regarding the ecological composition and size of the YUS Conservation Area in order to meet the needs of the species.

In June 2014, a research team including TKCP's Program Director Lisa Dabek, TKCP Research and Monitoring Coordinator Daniel Okena, field veterinarian Carol Esson, YUS Conservation Organization President Timmy Sowang, YUS Rangers James Jio and Geno Yuwoc, and several local landowners and field assistants spent two weeks at the research site to capture and radio-collar tree kangaroos. Three tree kangaroos were successfully captured and radio-collared, released, and have been tracked daily by the local research assistants through December 2014. To contribute to an ongoing health assessment of wild tree kangaroos, Dr. Esson and the research team also gathered morphometric data and biological samples for analysis.

From top to bottom:

TKCP Research and Monitoring Coordinator Daniel Okena (back row, center), YUS Rangers Geno Yuwoc (front, left) and James Jio (front, right), and tree kangaroo capture team. *Photo by Lisa Dabek, TKCP.*

Field veterinarian Carol Esson (front, center), Lisa Dabek, Daniel Okena, and research team with collared tree kangaroo. *Photo by TKCP.*

Field veterinarian Carol Esson examining collared tree kangaroo. *Photo by Lisa Dabek, TKCP.*

Right: YUS Rain forest. *Photo by Lisa Dabek, TKCP.*

Research Collaboration

TKCP partners with research institutions and scientists to answer key questions defined in the YUS Landscape Plan. Research findings can inform management decisions and guide the direction of the YUS Conservation Area. Also, TKCP and YUS CO facilitate outside scientists to conduct research in YUS. In 2014, YUS welcomed several research projects:

Research Project	Principle Investigator(s)	Institution
Matschie's tree kangaroo home range and habitat use	Lisa Dabek Daniel Okena YUS Tracking Team	Woodland Park Zoo TKCP-PNG
Matschie's tree kangaroo health assessments	Carol Esson Erika Crook Lisa Dabek Trish Watson Daniel Okena	Cairns, Australia Utah's Hogle Zoo Woodland Park Zoo TKCP-PNG
Evaluation of hunting sustainability and the potential role of protected areas in Papua New Guinea (PNG): Implications for conservation	Gabriel Porolak	James Cook University
Life histories of mid- to upper-montane robins (Petroicidae) endemic to New Guinea	Dr. Richard Donaghey	Independent Researcher
Language documentation and description of Tiatuk, the language of Gogiok	Dr. Valerie Guerin	James Cook University
Comparative thermal physiology and evolution of Australopapuan skinks: Testing hypotheses for gradients in species diversity	Charlotte Jennings	University of California, Berkeley

Future Plans

TKCP is always seeking opportunities to collaborate with researchers and scientists to investigate questions across the YUS landscape.

- Continue research on tree kangaroo home range and habitat use at differing elevations throughout the YUS CA

- Establish and apply an ecological monitoring system for marine species and ecosystems

- Study social issues relating to sustainable management of the YUS Conservation Area

- Investigate impacts of climate change in the YUS ecosystem

Planning for a Sustainable Future

Ward-level Land-Use Planning

Over the course of three years, TKCP, led by Conservation Strategies Manager Karau Kuna, has worked diligently with villages throughout YUS to develop land-use plans (LUP), in which landowners and communities zone their land and collectively define how their resources will be used. In 2014, TKCP and the people of YUS celebrated the completion of the final five of 18 ward-level land-use plans. Together these plans represent the commitment of 50 villages and over 12,000 people to the responsible management of their land and resources in the interest of conservation and sustainability for the YUS landscape.

The mosaic of the ward-level land-use maps conveys the significant impact that the plans will have in protecting biodiversity within and beyond the boundaries of the YUS Conservation Area. The detailed plans designate zones for specific conservation and development purposes according to the communities' self-defined priorities. Throughout the land-use planning process, additional landowners recognized the value of committing tracts of habitat for conservation or reforestation in

Throughout the land-use planning process, additional landowners recognized the value of committing tracts of habitat for conservation or reforestation in order to enhance ecosystem connectivity across the landscape.

order to enhance ecosystem connectivity across the landscape – as part of the 3,190 additional hectares pledged for conservation in 2014, a clan from Saburong village in ward 3 pledged 937 hectares of forest, which significantly improves ecosystem connectivity in the Uruwa region and connects the YUS research transect in the YUS CA.

This milestone completion of the land-use plans across YUS was widely celebrated. At a ceremony at Isan village in September 2014, the Governor of Morobe Province, Kabwum District's Member of Parliament, and the President of Yus LLG praised the communities for their commitment to conservation. As a symbol of our commitment to working with local landowners and government officials in pursuing the conservation and development goals identified by the people of YUS, TKCP ceremoniously handed the land-use plans to the representatives of Yus, Wasu and Nayudo LLGs.

 Morobe Provincial Governor Honorable Kelly Naru (right) and MP from Kabwum District, Honorable Bob Dadae (center) accepting YUS land-use plans from YUS CO President Timmy Sowang (left). *Photo by TKCP.*

STRUCTURE OF GOVERNANCE IN PAPUA NEW GUINEA

Village

Ward

Local Level Government

District

Province

Nation

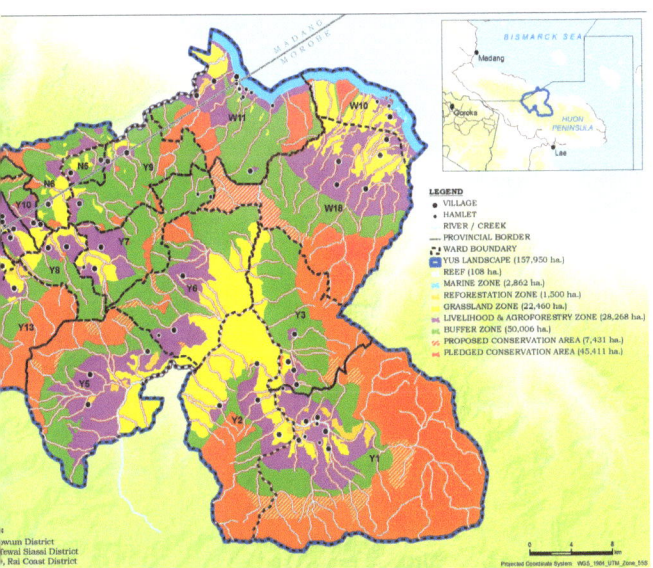

I thought everything was in order. I exploited my natural resources just as my father did, and his father before him. I paid little attention to what was actually left in the forest. After attending TKCP's first Land-use Planning workshop, I realized the depleted state of my resources and the importance of planning for their sustenance into the future. I now have the drive and the tools necessary to help me do that. In terms of government services, we don't have much but we utilize the few provisions available for our ward. I am happy my leaders have attended this workshop, and hope they can apply what they have learned to deliver much-needed services to our people. I am keen and will definitely be attending the next two LUP workshops. Thank you.

Danny James, Komareng village, Ward 5, Nayudo LLG, Rai Coast, Madang Province

From top to bottom:
TKCP's Conservation Strategies Manager Karau Kuna facilitating LUP workshop in Uruwa zone. *Photo by TKCP.*
Completed YUS landscape mosaic of ward-level land-use plans. *Map by Karau Kuna.*
YUS community members deliberate land zoning decisions. *Photo by TKCP.*

Land-use Planning: Local, National and Global Impact

- Because the community's needs and priorities for conservation, development, and resource management are addressed simultaneously, participants are able to identify creative solutions and compromises which relieve environmental pressures without limiting opportunities to improve community welfare and livelihoods. As a result, the communities of YUS are already committing additional land for conservation and reforestation.

- Illegal activities appear to be declining in the YUS Conservation Area, as people better understand the impact on the species they identified as priorities during the land-use planning workshops. Activities such as hunting and clearing are primarily limited to the designated zones created through the LUP process. Reports of an increased presence of once-scarce wildlife near villages and gardens are a promising indication of species population growth in the protected "wildlife bank" areas.

- The neighboring LLGs of Nayudo, Wasu, and Leron-Wantoat have expressed interest in establishing their own conservation areas, based on TKCP's land-use planning model in YUS. Conservation efforts in neighboring areas will further strengthen the conservation impact in YUS, as it mitigates threats along the boundaries of the land- and sea-scape. TKCP is honored to share our lessons learned and provide advice in support of these efforts.

- PNG's national Department of Environment and Conservation is interested in conducting land-use planning in other protected areas across the country. DEC has adapted TKCP's land-use zoning model for inclusion in the new PNG Protected Areas Policy, to be applied to protected areas throughout the country. The TKCP team is eager to share our experiences with the national government to support the creation of other protected areas throughout the country.

- The United Nations Development Program (UNDP) awarded TKCP-PNG with the Equator Prize – a highly-distinguished award recognizing local sustainable development solutions for people, nature and resilient communities. TKCP-PNG has been recognized as a "remarkable demonstration" of these qualities, and the land-use planning process noted as a Sustainable Forest Management model for community-led conservation throughout the globe. See Awards section, page 10 for more details.

Clockwise from top left:
Government representatives, YUS CO and TKCP celebrate the completed Land-use Plans and gazetted YUS bylaws at Isan village. *Photo by TKCP.*
YUS community members in traditional dress during YUS Landscape Plan celebration in Isan village. *Photo by TKCP.*
Gomdam village. *Photo by Lisa Dabek, TKCP.*
Karau Kuna at the United Nations Equator Prize award ceremony in New York, USA with fellow PNG prize winner Ursula Rakova of Tulele Paisa. *Photo by UNDP.*

Future Plans

- Develop community-based framework for monitoring effectiveness of ward-level land-use plans

- Increase community ownership of conservation through the use of participatory three-dimensional land-use mapping

Serving the Communities

Livelihoods

CONSERVATION COFFEE

TKCP's YUS Conservation Coffee initiative, in partnership with Caffe Vita in Seattle, Washington, continues to yield exciting results for farmers who have pledged land to the YUS Conservation Area. The project is entering its fourth year, and farmers in Uruwa zone exported another three tons of coffee beans to Caffe Vita – a coffee importer and roaster committed to socially and environmentally responsible coffee. Boosted by Caffe Vita's commitment to purchasing beans from YUS, TKCP has supported investment and farmer capacity-building, and has encouraged the development of government freight subsidies to improve farmers' access to markets. With additional potential from existing coffee trees throughout

Boosted by Caffe Vita's commitment to purchasing beans from YUS, TKCP has supported investment and farmer capacity-building, and has encouraged the development of government freight subsidies to improve farmers' access to markets.

YUS, the efforts continue to improve economic viability and increase productivity for the farmers in YUS.

Seeking to expand the coffee initiative beyond its pilot areas in Uruwa zone, TKCP's Ben Sipa worked together with PNG's Coffee Industry Corporation (CIC) and Caffe Vita's buyer, Daniel Shewmaker, to provide training and explore the feasibility of exporting coffee from villages in Yopno and Som zones. In October, Daniel joined a coffee husbandry workshop in YUS, committing with farmers to a goal of the first coffee exports from Yopno in 2015. Support by a new partnership between TKCP and Zoos Victoria in Melbourne, Australia, the program will continue to expand for the benefit of both the communities and wildlife of YUS.

 Clockwise from top left:
Coffee cherries. *Photo by Benjamin Sipa, TKCP.*
Caffe Vita's Daniel Shewmaker sorting coffee cherries with farmers in Yopno zone. *Photo by Benjamin Sipa, TKCP.*
TKCP Livelihoods and Community Services Manager Benjamin Sipa demonstrates use of a coffee pulper during a quality control workshop. *Photo by TKCP.*
Farmer taking notes during cocoa husbandry workshop. *Photo by Benjamin Sipa, TKCP.*
Cocoa pods growing in Nambis zone. *Photo by Benjamin Sipa, TKCP.*

As Caffe Vita's buyer I am filled with anticipation and excitement for the people of Yopno and their coffee. Having witnessed the positive impact in the Uruwa region, we are looking forward to replicating the same success in Yopno. Additionally, from a quality standpoint, the farmers of Yopno have the potential to produce some truly exquisite coffee that will be among the highest grown in the world. Our GPS read out elevations of 2,000-2,400 meters above sea level at all of the farms we visited. With massive swings in temperature between day and night, ridiculously fertile black soils, and old growth heirloom varieties of coffee, the sky is the limit.

Daniel Shewmaker, buyer for Caffe Vita
www.caffevita.com/blog/2014/10/yopno-coffee-training/

We never got such prices before the conservation coffee program. Families who have kids in high school see this as an opportunity to offset the school fees from what we earn. This is such a big relief, in addition to the benefits for conservation. I can see parents now who are able to afford new clothes for their kids to wear to school. And families can even earn enough to keep a small savings which are invested into equipment and materials for next year's harvest.

Annie Ogate, Yawan Village, mother, coffee farmer, YUS CO executive member, and elementary school teacher

CONSERVATION COCOA

With support from UNDP's GEF Small Grants Programme and Vibrant Village Foundation, TKCP aims to replicate the success of the YUS Conservation Coffee initiative among the cocoa-producing villages along the coast of YUS. TKCP's Community Services and Livelihoods Manager Ben Sipa provided farmers, who have pledged land for conservation, with capacity-building opportunities in husbandry and quality management techniques. These included an exchange visit with The Nature Conservancy and Adelbert Cocoa Cooperative program in neighboring Madang Province. The Adelbert Cooperative is one of the only groups in PNG working to establish protected areas through the production of environmentally-friendly fair-trade cocoa, and the exchange visit offered valuable insights into production, transport, and industry certification for YUS Conservation Cocoa.

In 2014, YUS cocoa farmers supplied more than 550 kg of fermented cocoa beans to PNG-based Paradise Foods Limited for its Queen Emma Chocolates brand, receiving a 30% premium above local market rates. The boutique chocolatier is committed to Zsupporting TKCP's conservation efforts with a portion of their income from YUS Kakao chocolate sales.

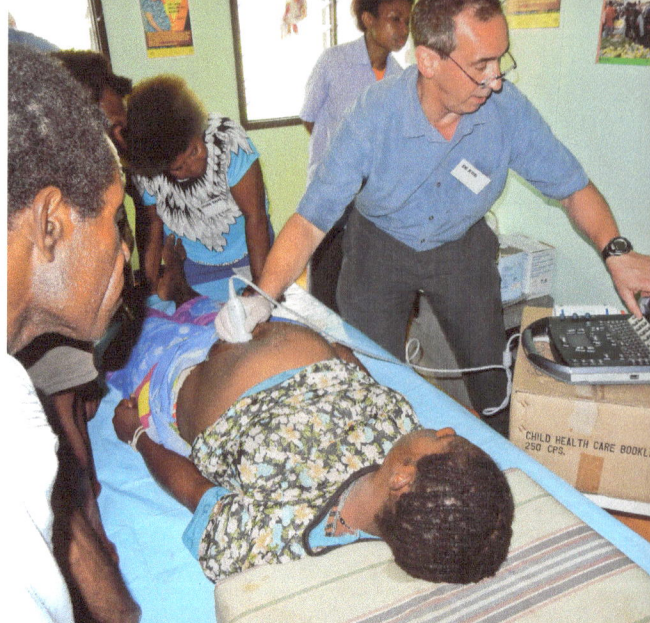

Health

HEALTHY VILLAGE, HEALTHY FOREST

Through a new partnership with Vibrant Village Foundation, TKCP welcomed a Healthy Community Coordinator, Melanie Palili, to lead the program's efforts to address the health needs of communities in YUS. In October, TKCP and Hesing Wayain organized a Health Training Workshop in Sapmanga village, aiming to advance the knowledge and skills of YUS Community Health Workers, Health Aids and Village Birth Attendants (midwives). Facilitated by five volunteer physicians from the United States, the workshop focused on issues relating to pregnancy, childbirth and nutrition, as well as common illnesses including malaria and respiratory issues.

YUS Health Visit Team:

- Dr. Robert Liddell, Diagnostic Radiologist, Seattle, Washington

- Dr. Marti Liddell, MD, Internal Medicine Physician at Polyclinic, Seattle, Washington

- Dr. Nancy Philips, Emergency Medicine Physician at Dartmouth-Hitchcock Medical Center, Lebanon, New Hampshire

- Dr. Blair Brooks, Internal Medicine Physician at Dartmouth-Hitchcock Medical Center, Lebanon, New Hampshire

- Dr. Travis Austin, Emergency Medicine Resident at Dartmouth-Hitchcock Medical Center, Lebanon, New Hampshire

- Sister Baleb Wahazoka, former Family Health Services Coordinator at Morobe Provincial Health Department

- Melanie Palili, TKCP Healthy Community Coordinator

- Daniel Okena, TKCP Research and Monitoring Coordinator

- Dr. Lisa Dabek, TKCP Program Director

I wanted to learn more and improve my knowledge to care for patients better. This workshop has broadened my knowledge and I plan to share this new knowledge with others so that together we can provide better health care services.

Hesing Wayain, Community Health Worker and Co-coordinator of the YUS Health Training Workshop

Clockwise from top left:
Sister Balob Wahazoka and Dr. Blair Brooks speaking at health workshop. *Photo by TKCP.*
Dr. Rob Liddell using an ultrasound to give Village Birth Attendants their first glimpse of a fetus in the womb. *Photo by Lisa Dabek, TKCP.*
Health workshop team and participants at Sapmanga village. *Photo by Daniel Okena, TKCP.*
Dr. Nancy Philips (front, left) leads a discussion with female Village Birth Attendants. *Photo by Lisa Dabek, TKCP.*
TKCP Healthy Communities Coordinator Melanie Palili (left) translates while Dr. Travis Austin (right) and Dr. Marti Liddell (second from right) see patients in Sapmanga village Health Clinic. *Photo by Lisa Dabek, TKCP.*
Dr. Rob Liddell, Dr. Blair Brooks, Dr. Travis Austin, and Daniel Okena speak with Hesing Wayain, male Community Health Workers and Village Birth Attendants. *Photo by Lisa Dabek, TKCP.*

Since completing my grade six education, I returned to my village and have taken on the role of a Village Birth Attendant (VBA). My wife is also a VBA and I have seen that her work has helped many women—therefore I decided that I should be a VBA as well, to assist my wife and other women in the village. I have been a VBA for eight years, even though I have had no proper training. This workshop has taught me new skills such as removing the umbilical cord from around the baby's neck, treating joint pains, and family planning.

Jack Miyaning, Village Birth Attendant

Participants learned techniques for diagnosing and treating basic health problems, and gained an understanding of established patient referral systems for the treatment of serious illnesses and injuries. Using a portable ultrasound machine, Dr. Rob Liddell provided the participants with a thrilling first opportunity for local village midwives and healthcare workers to witness a fetus developing in a pregnant mother's womb. Basic medical equipment and supplies were provided by the visiting physicians and by OilSearch Women's Network for use in health facilities throughout YUS, including weight scales, stethoscopes and informational posters.

Following the workshop, the visiting doctors attended to over 200 patients who had travelled from throughout the YUS landscape for the rare opportunity to consult with a doctor. Among the many cases, three patients were referred for urgent lifesaving care at the Angau Provincial Hospital in Lae. TKCP team and the visiting physicians also met with representatives and doctors from the Provincial Health Department, Lutheran Health Services and Etep rural hospital, laying a foundation for patient referrals and treatment, as well as opportunities to collaborate with strengthening the health systems throughout YUS. TKCP appreciates the health team for their dedication to this initiative.

Education and Capacity-Building

BALOB TEACHER'S COLLEGE SCHOLARSHIP

YUS community members often cite TKCP's teacher scholarship program as essential in ensuring the next generation of environmentally-aware citizens are educated. For over a decade, TKCP has helped to fill vacant teaching positions in YUS with qualified teachers through a scholarship program. Since 2002 TKCP has helped 22 YUS students to graduate as teachers who return to teach children in YUS. Currently, 16 graduates are now teaching in YUS schools. Four additional teachers supported by TKCP have been promoted to senior positions throughout Morobe Province. TKCP congratulates Mr. Elizah Bowong of Kalaset village and Mr. Horangke Muyo of Urop village, as they graduated from Balob Teachers College in Lae in 2014 and have accepted teaching positions in YUS commencing in February 2015.

JUNIOR RANGERS

The TKCP Junior Ranger Program returned to YUS in 2014 after a short hiatus. Aiming to connect youth with the wonders of nature around them and to encourage conservation actions, the program welcomed 27 elementary students from Westkokop village. The Junior Rangers participated in YUS Environment Day, sharing their new understanding of conservation and the environment, and demonstrating how everyone can do their part to ensure a sustainable environment into the future. The Junior Ranger Program will continue to develop in 2015.

Clockwise from top left:

Balob Teachers College Scholarship recipients Mr. Elizah Bowong (left) and Mr. Horangke Muyo at graduation ceremony. *Photo by TKCP.*

TKCP Leadership Training and Outreach Senior Coordinator Danny Samandingke. *Photo by Mikal Nolan, TKCP.*

Girls performing a singsing dance in Sapmanga village. *Photo by Lisa Dabek, TKCP.*

Children at the Sapmanga village school. *Photo by Nancy Philips.*

Junior Conservation Rangers at Westkokop village. *Photo by Danny Samandingke, TKCP.*

Primary school chalkboard in Saburong village. *Photo by Nancy Philips.*

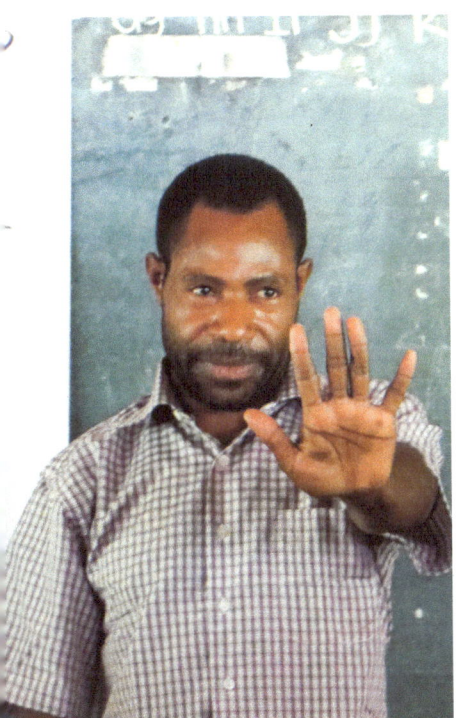

LEADERSHIP TRAININGS

To ensure the long-term sustainability of the conservation and community development efforts throughout YUS, TKCP strives to inspire and strengthen local YUS leaders who will educate and rally their communities to action. Led by Leadership Training and Outreach Senior Coordinator Danny Samandingke, the program engages local participants to provide trainings based on traditional customs and management concepts while incorporating modern methodologies. Samandingke provided training to 98 village leaders and community members participating in TKCP's livelihoods and Land-use Planning initiatives, which will further enable them to lead, organize, and manage their plans for conserving the forest and improving their livelihoods.

The leadership training has enhanced my knowledge and understanding of how I can effectively communicate and serve as a community peace-making officer. I understand the importance of creating a clear vision, creating a community calendar and a plan, establishing a strong community leadership circle, and applying simple management steps. I also realize that following a plan is essentially important too, especially when dealing with fines and charges (in the YUS CA).

Mr. Jack Samboki, Community Peace-Making Officer

YUS Community Organization

The YUS Conservation Organization (YUS CO), a community-based organization comprising of 21 men and women from across the YUS landscape, has evolved into a strategic partner for TKCP. It is an organization of landowners who advocate for community initiatives and advise TKCP on YUS community needs. Together the organizations seek to encourage appropriate management decisions for YUS CA, and endeavor to meet the needs of people, communities and environment.

In 2014, YUS CO President Timmy Sowang was appointed by Yus LLG President Hon. Epemu Kiwenu to be the LLG's first conservation representative. Timmy traveled to Port Moresby to offer suggestions gathered on behalf of YUS landowners for the proposed draft of the new PNG Protected Areas Policy. During the visit, he also had the opportunity to share practical insights from the YUS Conservation Area with UNDP consultants advising on the development of the new policy.

 Clockwise from top right:

Members of YUS CO and TKCP on a learning exchange visit to a community nursery project at Marawasa village in the Markham Valley. Photo by TKCP.

Woman in traditional dress at Isan village. Photo by Mikal Nolan, TKCP.

YUS CO Executive member Annie Ogate leads group discussion. Photo by TKCP.

YUS CO Executive members and the YUS CO annual general meeting organizing committee members. Photo by Mikal Nolan, TKCP.

KUTIM NA PLANTIM (CUT AND PLANT) COMMUNITY CONSERVATION PROGRAM

2014 marked a milestone for YUS CO as they managed their own activities for their "Kutim na Plantim" project, funded by UNDP's GEF Small Grants Programme. The YUS CO Executive team worked with faculty from the PNG University of Technology's (Unitech) Department of Forestry and the Forest Research Institute (FRI) to conduct a week-long training-of-trainers focusing on the development of community forest nurseries. The nursery project aims to produce locally and culturally important trees to reduce harvesting pressures on the protected areas, while ensuring these trees are available for future generations. On World Environment Day, YUS CO organized a celebration to promote the initiative and encourage people to plant five trees for every one tree harvested. In honor of the YUS culture and environment, elementary students performed drama shows and sing-sings.

Future Plans

- Expand conservation coffee project into communities of Yopno and Som zones

- Establish quality standards and transport logistics for conservation cocoa initiative

- Develop strategic partnerships and plans for a Healthy Communities initiative in YUS

- Support farmers to establish the YUS Conservation Coffee and Cocoa Cooperative Society

- Continue supporting the organizational capacities of YUS CO

- Build the YUS Junior Ranger program

Operating TKCP

Learning and Sharing Our Work

TKCP is always searching for new ideas and methods to improve our work, and eager to share our lessons for the benefit of others. The team had many opportunities to connect with partners and colleagues, both in Papua New Guinea and at conferences around the world. With support through the Pacific Island Countries Participation Fund (PIC Fund) managed by the New Zealand Aid Programme, most of the TKCP leadership team traveled to Sydney, Australia in November 2014 to participate in the IUCN's World Parks Congress. The event is held every 10 years and drew more than 6,000 attendees from over 170 countries representing the public and private sectors to explore current issues on protected areas and approaches for conservation and sustainable development around the globe. Each member of the team presented elements of TKCP's work:

- Benjamin Sipa highlighted TKCP's experiences with Conservation Coffee and Cocoa.

- Karau Kuna presented on the innovative community-led Land-use Planning process and how it contributes to community resilience.

- Lisa Dabek was invited by IUCN and Save Our Species (SOS) to present the program on two occasions during the event, including a discussion on the important role of small grants in supporting conservation efforts. She also presented on the value of financial sustainability for conservation through the YUS Endowment.

- Mikal Nolan shared about the value of community ownership and TKCP's experiences from the BALANCED Project on a panel exploring the linkages between human health and environmental conservation.

- Danny Samandingke presented TKCP's approach to building and empowering young leaders.

- Daniel Okena shared an overview of endangered species in YUS and connected with the World Federation of Rangers, enabling him to link TKCP's YUS Conservation Ranger program with others around the world.

The TKCP team also connected with other UNDP Equator Prize winners as part of a new WIN Pacific network. The World Parks Congress was an excellent opportunity for TKCP to share our methods and lessons learned, gain new insights and ideas to build upon our efforts, and connect with a wide range of current and future partners.

The Asian Pacific Rainforest Summit and Stakeholder Dialogue took place in Sydney, Australia before the World Parks Congress in November, bringing together professionals from governments, non-government organizations and corporations in the region. Mikal Nolan, TKCP's Program Manager, attended panel discussions on the topics of managing forests, protecting biodiversity and involving indigenous communities.

 Members of TKCP leadership team at World Parks Congress in Sydney, Australia. (from left) Danny Samandingke, Benjamin Sipa, Lisa Dabek, Karau Kuna, Mikal Nolan Daniel Okena. *Photo by Sandra Elvin, AZA.*
Karau Kuna speaking on a panel at the 2014 World Parks Congress. *Photo by TKCP.*

With Karau Kuna's appointment to the Morobe Province Climate Change Committee and expertise in facilitating community-led Land-use Planning, he was invited by PNG's National Department of Environment and Conservation (DEC) to participate in the second Protected Areas Policy Development Workshop. As the only nationally-gazetted Conservation Area in PNG, TKCP is considered a model for protected areas in the country. Based on the success of Land-use Planning among the communities of YUS, DEC has incorporated the process into the draft PNG Protected Areas Policy. When approved, the Policy will guide the management and governance of nationally-protected areas throughout the country.

In June, Mikal joined the Morobe Province delegation at a workshop sponsored by PNG's Office of Climate Change and Development and UN-REDD. The meeting highlighted the newly-developed REDD+ training manual, and generated discussions to shape future policies and implementation of REDD+ Safeguards and Free, Prior, and Informed Consent for PNG landowners and communities.

Danny Samandingke was invited by Seattle's iLEAP leadership program to return to Seattle as one of two alumni to address over 100 people at a fundraising event. He spoke of the iLEAP training experience, and how he has applied his leadership knowledge to his work with communities in YUS. Danny has integrated the tools and concepts he learned through the program with traditional concepts to develop the leadership capacity of landowners and community members throughout YUS. "The connections established through iLEAP have been truly amazing tools, molding my professional goals to serve with a missionary's heart of a servant, and I respect iLEAP for developing that. Lest I forget my college motto, 'Balob To Serve.'" Danny was grateful for the opportunity to share his experience and passion with iLEAP fellows and supporters, and built new relationships with grassroots leaders. During his time in Seattle, Danny and Lisa Dabek also met with members of the Lummi Nation in northern Washington state to share perspectives on indigenous knowledge and community conservation in the hopes of creating a learning exchange program in the future.

"The connections established through iLEAP have been truly amazing tools, molding my professional goals to serve with a missionary's heart of a servant..."

Danny Samandingke speaking at iLEAP's fundraising event in Seattle, USA. *Photo by iLEAP.*

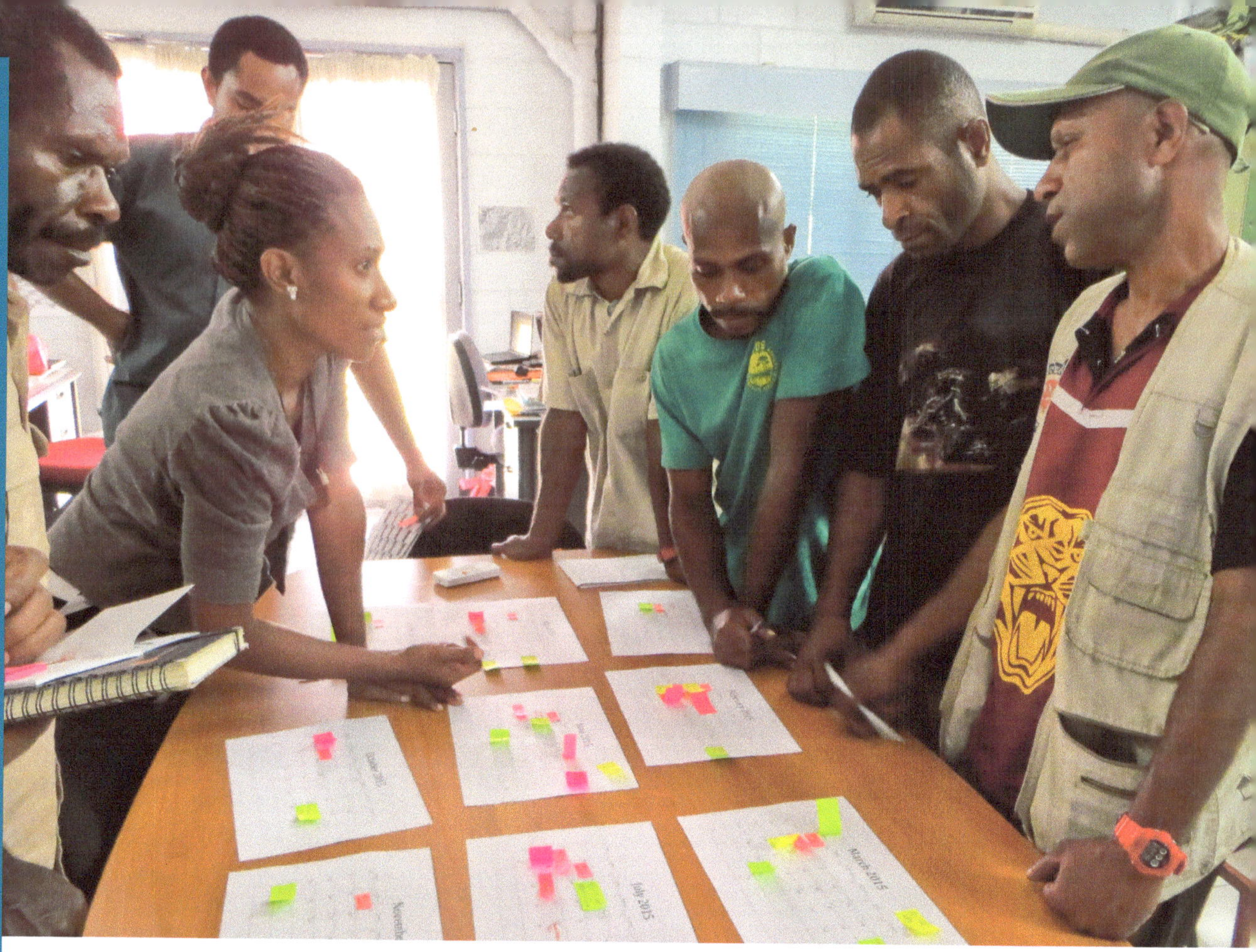
March 2015
July 2015

World Environment
theme:

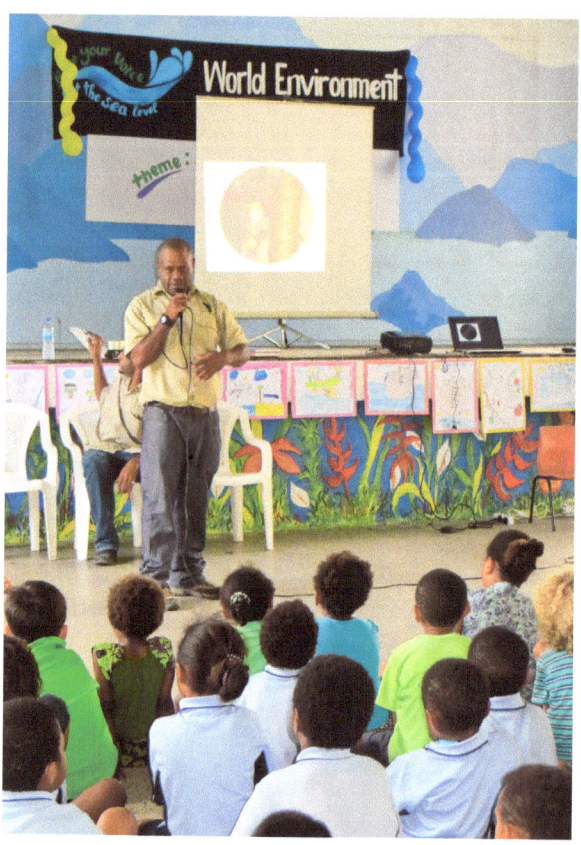

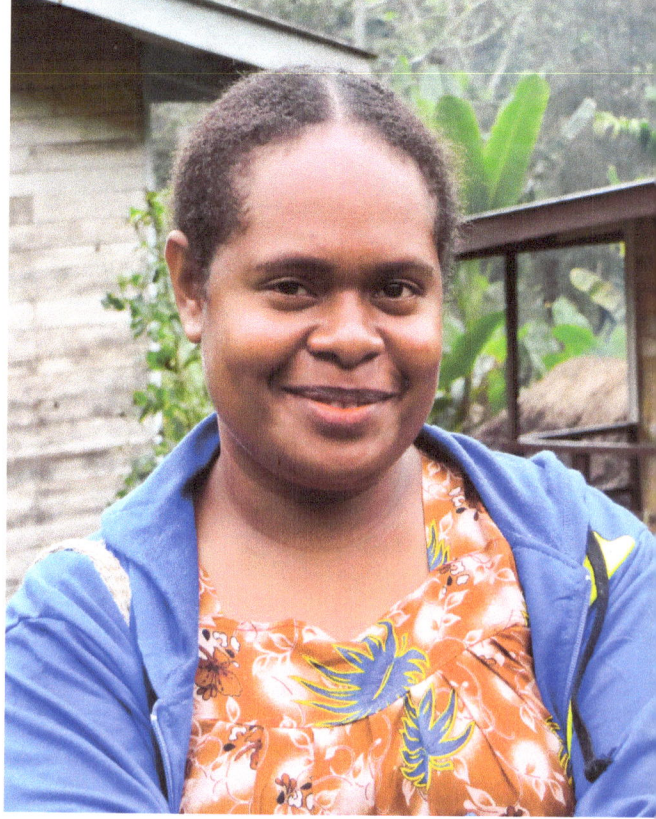

Clockwise from top left:

TKCP team discussing program plans in the Lae office. *Photo by Mikal Nolan, TKCP.*

TKCP Program Manager Mikal Nolan (second from right) trains Conservation Officers to use computers to manage YUS CA Ranger monitoring data. *Photo by Lisa Dabek, TKCP.*

TKCP Administrative Coordinator Laris Bartsaka. *Photo by Mikal Nolan, TKCP.*

Benjamin Sipa speaks to students at the Lae International School on World Environment Day. *Photo by TKCP.*

Professional and Organizational Development

Following the establishment of our local non-governmental organization in 2013, TKCP-PNG assumed greater independence and responsibility in 2014 for the management and administration of the program. To provide direction and shape the continued development of TKCP-PNG, a Board of Directors has been appointed:

- Chair: Dr. Lisa Dabek (Woodland Park Zoo)
- Dr. Bruce Beehler (Smithsonian Institution)
- Mr. Francis Hurahura (The Nature Conservancy PNG)
- Mr. Zachary Wells (Conservation International)
- Ms. Mikal Nolan, ex officio (TKCP-PNG)

Managing its first local grant from UNDP GEF Small Grants Programme, the organization continues to demonstrate its capacity to implement TKCP's work effectively and transparently. Laris Bartsaka is in charge of TKCP-PNG's financial tracking. Danny Samandingke and Lisa Dabek participated in a First Aid and CPR class at WPZ taught by WPZ's Trinidad Alcaraz. WPZ continues to provide support and oversight for the further development of TKCP-PNG's systems and processes.

In March, TKCP-PNG conducted a Monitoring and Evaluation Workshop with the team in Lae, led by capacity-builder Maureen Knijff with Voluntary Service

Overseas. As TKCP strives to measure and communicate the impact of its initiatives, the team gained a broader understanding of the purpose, processes, and tools used in monitoring and evaluating program effectiveness. With plans to establish and strengthen monitoring systems at both the program- and community-level in 2015 and beyond, the team was very excited to better understand the framework and steps in developing a monitoring system.

Additionally, to improve upon the value and accuracy of the YUS Conservation Area monitoring data collected by the Ranger team, TKCP provided YUS Rangers with training in the use of GPS devices and waypoints, and equipped the four Conservation Officers with field laptops and computer training for entering and transferring data to the Lae office for further analysis. The new tools and knowledge will be helpful for the team, saving time and allowing a faster response to trends and issues identified through the data analysis.

Future Plans

- Continue to strengthen staff and organizational capacity of TKCP-PNG
- Continue to build the YUS Conservation Endowment
- Update and reconfirm YUS Landscape Plan for 2016-2020 to correspond with government planning cycles
- Continue to share lessons learned at workshops and conferences

TKCP in the Media

TKCP's efforts and accomplishments were highlighted in both Papua New Guinea newspapers and in media around the world in 2014.

UNDP EQUATOR PRIZE

The UNDP Equator Prize earned TKCP attention and praise both locally and internationally. PNG's national newspapers, the Post Courier and The National, featured our work in several articles celebrating the award and its importance for Papua New Guinea. PNG's leading online news network, PNGLoop.com, also featured TKCP's award on their website. We are honored to see the conservation commitment of YUS communities recognized throughout the country.

1. PNG Post Courier article, February 5, 2015
2. PNG Post Courier article, November 7, 2014
3. PNG Post Courier article, November 21, 2014
4. pngloop article: www.pngloop.com/2015/02/04/community-led-projects-win-equator-prize-award/

The United States Embassy and United Nations Development Programme in Papua New Guinea also celebrated TKCP's Equator Prize and Special Recognition Award. In June, Lisa Dabek and Mikal Nolan met with U.S. Ambassador Walter North and USAID Pacific Islands Regional Director Maurice Knight to discuss the program and its accomplishments.

5. Article: portmoresby.usembassy.gov
6. Article: www.pg.undp.org

Internationally, the IUCN recognized TKCP's award, and our work was featured prominently through the UNDP Equator Initiative's announcement of prize winners. This attention raises our profile among conservation programs throughout the world, helping to expand our network and our learning.

7. Article: www.iucn.org
8. Article: equatorinitiative.org

TKCP in the Media

MILESTONES AND ACCOMPLISHMENTS

2014 was a very exciting year in YUS! The completion of TKCP's Land-use Planning was featured in two articles in PNG's The National newspaper, highlighting the value of the plans in guiding government resources and investment. In support of the YUS Landscape Plan, Local and Provincial government representatives pledged funding for community-led initiatives.

We are honored that the IUCN and Save Our Species program shared the impacts of TKCP's Conservation Coffee project in a widely distributed article, "Tree Kangaroo Friendly Coffee – A Triple-Win Story." We are excited to share our lessons and successes to help other organizations and projects around the world.

NATION

Morobe Governor Kelly Naru (left) and Kabwum MP Bob Dadae and at Isan village last Sunday.

Morobe 'to diversify'

By PISAI GUMAR

MOROBE Governor Kelly Naru is determined to push for development activities in infrastructure, agriculture, mining and protection of bio-diversity in the province.

He said he did not want to push for the same projects in the nine districts, especially roads, gas and mining in areas where bio-diversity activities had already affected the people.

"The Morobe provincial government is mindful never to develop and exploit all its resources at once," he said.

"If Bulolo and parts of Huon Gulf are involved in mining activities, then Markham should become the agricultural food bowl.

"Menyanya, Wantoat, Kabwum and interior Finschhafen can produce coffee and fresh foods to supply the urban markets.

"Coastal communities in Huon

Gulf, Finschhafen, Tewae-Siassi and Nawaeb can supply marine produce, including cocoa, taro and banana."

He said districts must identify and plan their resources development and the provincial government would come in to support them financially.

Naru was at Isan village in Yus council, Kabwum, last Sunday after the Tree Kangaroo conservation programme received the United Nations Development Programme equator prize.

"Road is an essential part of progress but must be properly surveyed and constructed without trespassing the conservation areas," Naru said.

Meanwhile, MP Bob Dadae will build a road to link Yus to Wantoat in Markham and told the people to properly identify the route so that it does not infringe on protected areas.

Dadae gave K1 million for two communication towers in the area.

NATION

Naru (from right), Yus council president Efemuc Kiwenu, Kabwum MP Bob Dadae and TKCP-PNG programme manager Mikal Eversole Nolan displaying the integrated ward development and land use planning in Yus landscape at Isan village last Sunday. Note a villager (front) in a kangaroo dress.

Tree kangaroo plan gets fund

POLITICAL leaders in Morobe have pledged K300,000 to support the Tree Kangaroo Conservation Project (TKCP) in the Yus council area in Kabwum.

The leaders included Governor Kelly Naru (K150,000), Kabwum MP Bob Dadae (K100,000) and Yus council president Efemuc Kiwenu (K50,000).

Their financial backing was prompted after United Nations Development Programme (UNDP) Equator Initiative announced TKCP-PNG as a winner of the Equator Prize 2014.

The award followed a global call for nominations for the Equator Initiative that received a record 1,234 nominations from

121 countries around the world.

The Technical Advisory Committee of international experts guided the review and selected 35 winners, representing outstanding local achievement in advancing sustainable development solutions for people, nature and resilient communities.

While diverse in their innovations and areas of work, each winner demonstrated community-based grassroots action to address environment, poverty and climate change challenges head-on.

The leaders unveiled to Yopno II communities in Isan village last Sunday the integrated ward development and land use plan-

ning in Yus landscape, Isan circuit centre, parish office, Sinai Gentamo School, water supply and a primary school through an AusAID building programme.

TKCP-PNG programme manager Mikal Eversole Nolan acknowledged the promises from leaders to encourage and promote conservation activities in the Yus area.

Nolan said that government support was timely, after almost 20 years of commitment in conservation of various species of flora and fauna.

"TKCP-PNG is proud to work in partnership with Morobe provincial government, Kabwum MP and the Government."

9.   The National article, October 1, 2014

10.   The National article, October 1, 2014

11.   Article: www.sospecies.org/sos_news/success_stories

12.   Article: www.iucnredlist.org/news/tree-kangaroo-friendly-coffee-a-triple-win-story

The YUS Conservation Coffee project was also highlighted by TKCP's new supporter, Zoos Victoria, in the December issue of "Zooper Kids" magazine. We are proud to share our work with future conservationists around the globe.

Our partners and visitors are also helping to get the word out about our work with the communities of YUS! Caffe Vita's Daniel Shewmaker shared his experiences and optimism about YUS Conservation Coffee following his visit, and Dr. Marti Liddell recounted her time in YUS training health workers and treating patients.

SAVING THE MATSCHIE'S TREE-KANGAROO

Zoos Victoria has teamed up with Woodland Park Zoo in Seattle, USA plus Papua New Guinea communities and the Papua New Guinea Government to support the Tree Kangaroo Conservation Program. The program is working to save a particular species of tree-kangaroo, called Matschie's Tree-kangaroo.

The island of New Guinea has the third-largest area of tropical forest in the world, so the forests there are hugely important. A lot of the plants and animals there exist nowhere else in the world – including the Matschie's tree-kangaroo.

Mining and logging, high levels of poverty, limited access to health and education, and new jobs that don't impact on the environment are some of the challenges in Papua New Guinea.

The Tree-Kangaroo Conservation Program has worked with local communities to create the first-ever Conservation Area in Papua New Guinea, called YUS. It's named after three local communities, which are called Yopno, Uruwa and Som. The communities are working together to protect the land and animals.

As well as the Conservation Areas, which are patrolled by rangers, a Conservation Coffee Project has been created. The coffee project encourages farmers ... to protect land for tree ... in return, they're helped ... from different countries ... how to grow great ... use of chemicals.

MATSCHIE'S TREE-KANGAROO FACTS:

- Live only in the Huon Peninsula in eastern Papua New Guinea
- Endangered
- Spend most of their time in trees
- Eat mostly leaves but can also eat flowers, grass shoots, moss and bark
- Can jump from tree to tree
- Are mostly orange or orange-brown, with yellow feet, ears and belly
- Females have a pouch
- Joeys leave their mothers at about 18 months old

MEET THEMBI THE GIRAFFE

THE POLYCLINIC
News & Blog

Dr. Marti Liddell returns from volunteering in Papua New Guinea

By: Liddell, Marti Hyatt, MD | December 23, 2014

CATEGORY ARCHIVES: PAPUA NEW GUINEA

YOPNO COFFEE TRAINING | October 17, 2014

The YUS Conservation Area includes three distinct regions, named for the main rivers and languages spoken – Yopno, Uruwa, and Som. It was four years ago that we began to work with the Tree Kangaroo Conservation Program (TKCP) in developing a sustainable coffee program as part of their mission to improve the livelihoods of those who have pledged their land for conservation. The pilot project began in Uruwa, which had the most coffee and also were furthest along in terms of management and processing know how. Having just completed our fourth harvest with the farmers of Uruwa, we have turned our focus to Yopno.

Support

The Tree Kangaroo Conservation Program is extremely fortunate to have a close network of dedicated long-term supporters in Seattle and beyond. Our extended TKCP family members generously contribute their time, expertise, and resources in support of the YUS community, serving as advocates for tree kangaroos and for global wildlife conservation.

During TKCP Program Manager Mikal Nolan's visit to Seattle, Drs. Rob and Marti Liddell welcomed some of the TKCP family to their home for an evening of celebration and gratitude. The guests listened to Mikal's updates of the progress and accomplishments in YUS and shared their stories of visits to Papua New Guinea, their memories throughout TKCP's history, and their excitement for the program and the people of YUS. We are thankful to the Liddells for hosting this wonderful event, and look forward to meeting TKCP family members again.

TKCP supporters celebrating the program's milestones in Seattle, USA. Front row, from left: Maria Semple, Lisa Dabek, Poppy Meyer; Middle row, from left: Laura Baumwell (WPZ), Margie Wetherald, Cammi Libby, Jan Trasen, Mikal Nolan, Cyndi Wolfe, Marti Liddell; Back row, from left: Rob Liddell, Stanley Leahy, George Meyer, Gay Jensen, Len Barson, Trevor Holbrook, Kevin Schofield, Bob Plotnick, Fred Koontz (WPZ), John Lee, Sheri Flies. *Photo by WPZ.*

TKCP supporter Pascal Blumenthal (left) with Lisa Dabek. *Photo by WPZ.*

TKCP and Woodland Park Zoo are proud to encourage our younger generations to appreciate the natural world and to take action for wildlife conservation. This year we would like to acknowledge these Young Conservationists, including 5-year-old Penelope Guist and 15-year-old Pascal Blumenthal. It means a lot for our program to capture the interest of these young supporters, and we are thrilled to share in their passion for tree kangaroos. We had the opportunity to speak with Pascal to learn about his first tree kangaroo encounter, and how the experience has inspired him into taking action.

"I remember going to the zoo and seeing tree kangaroos for the first time when I was seven. They're so bizarre, and the most bizarre things always stand out. After I learned about them, I wanted to help. We set up a lemonade stand during a couple of days in the summer. We called it the 'Tree Wallaroo Lemonade Stand', and we had a big picture of a tree kangaroo. Everyone was charmed by it.

It's hard to expect someone to feel empathy for tree kangaroos if they've never gotten exposure to nature besides the squirrels in their neighborhood. And I love that for people who can't go hiking or camping often, the zoo is the gateway. It was the gateway for me, and for my interest in tree kangaroos. I wouldn't have known they existed without the zoo.

There's a gap between wanting to take action, and actually doing so, where I see a lot of people getting stuck. And the great thing about organizations like TKCP is that they provide that gateway, they appreciate your efforts no matter how small they are, and on a very personal level. They make you feel like you're part of the team, and that's a wonderful experience."

Thank you Pascal!

TKCP depends on a large network of people and organizations to help us accomplish our work in YUS. Many of our supporters are long-term partners and friends, and we sincerely thank you for your dedication and contributions. Together, we have achieved a great deal of success. We wish to thank our donors throughout the years and thank the following friends of TKCP for their financial and in-kind support in 2014.

$50,000 – $99,999

Conservation International

IUCN Save Our Species

Lifeweb Initiative of the German Federal Ministry of Environment (BMU) and German Development Bank (KfW) through Conservation International

UNDP GEF Small Grants Programme

The Vibrant Village Foundation

Zoos Victoria

$10,000 – $49,999

Anonymous

Columbus Zoo and Aquarium

Merrick and Lorraine Darley

Rob and Marti Liddell

Mohamed bin Zayed Species Conservation Fund

Nancy Philips and Blair Brooks

The Shared Earth Foundation

Cyndi Wolfe

$5,000 – $9,999

CREOI

Detroit Zoological Society

Enlyst Fund

Anne Mize and the Muchnic Foundation

New England Biolabs Foundation

New Zealand Aid Programme

Texas Instruments Foundation Matching Gifts

$1,000 – $4,999

Albuquerque BioPark

Cleveland Zoological Society/Cleveland Zoo

Milwaukee County Zoo

Richard and Ginger Goldman

Ruth and Terry Lipscomb

Sedgwick County Zoo

ZooParc de Beauval

Up to $999

Anonymous

Harriet Allen, in memory of Dr. Holly Reed and Ms. Shelby

Glen and Susan Beebe

Greater Kansas City American Association of Zoo Keepers

Pascal Blumenthal

The Boeing Company Matching Gifts

John and Sarah Brooks

G. Stephen and Diana Crane

Dominic and Winnie Deguzman

Janice and William Fischel

Ben Flaumenhaft

Carl and Nancy Guist

Derek, Emily, and Penelope Guist

Jonathan Guist

Meredith and John Jendzurski, on behalf of Joanna and Mary Elena

Mike Kaputa and Suzanne Tomassi

Summer Lopez

Microsoft Corporation Matching Gifts

Davi Ann and Matt Norsworthy

Oaklawn Farm Zoo

Rohrbach Family

William Schneider

Gena Shurtleff

Gary Smith and Christine May

Anne Stein

Amanda Tumulty

Jacob and Kristina Werenko

Andrew Wu

Support: YUS Conservation Endowment

TKCP and WPZ are proud to have established the YUS Conservation Endowment with Conservation International, which provides annual funding support for the management and protection of YUS CA in perpetuity. TKCP is grateful to all of our supporters who have helped us invest over $2 million for the creation of the YUS Conservation Endowment. We thank the WPZ Board of Directors for managing the endowment. The investment continues to grow, and since 2013 has supported 20 percent of TKCP's annual operating expenses in Papua New Guinea!

$100,000 and above

Anonymous

Conservation International

Estate of Lorene E. Currier

$50,000 – $99,999

Cammi and Jeff Libby

George Meyer and Maria Semple

Robert Plotnick and Gay Jensen

John F. Swift

Swift Family Fund

Margie Wetherald and Len Barson

$20,000 – $49,999

Blumenthal-Edsforth Family

Columbus Zoo and Aquarium

Nina Dabek and Peggy Shannon

Microsoft Corporation

Roger Williams Park Zoo

Kevin M. Schofield

The Shared Earth Foundation

Susie Wyckoff

$5,000 – $19,999

Anonymous

Paul and Sarah Balle

Sonya and Tom Campion

Merrick and Lorraine Darley

Stuart N. DeSpain

Serena and Neal Friedman

Lynn Hall*

Ted and Tara Hart

Rosemarie Havranek and Nathan Myhrvold

The Hoffmann Family

Rampa Hormel, Enlyst Fund

Carol and Bruce Hosford

Leonard and Norma Klorfine

Stuart Klorfine

Klorfine Foundation

Victoria Leslie

Trish Miner

Richard Saada

Sedgwick County Zoo

Maryanne Tagney and David T. Jones

Craig Tall

Gail Warren

$1,000 – $4,999

Anonymous

John and Andrea Adams

Adobe Systems, Inc. Matching Gifts

Albuquerque BioPark

Jane Alexander and Edwin Sherin

Paul Balle

Anthony and Lillian Bay

Glen and Susan Beebe

Laura Bentley

David Brunelle

Mark Christiansen

Cleveland Zoological Society/Cleveland Zoo

Michael and Lois Craig

Richard and Ginger Goldman

IBM Corporation Matching Gifts

Sugi Kana

Glenn Kawasaki

Rob and Marti Liddell

Ruth and Terry Lipscomb

Bert and Susan Loosmore

Macbeth Family

Milwaukee County Zoo

Daniel and Meredith Morris

 Clockwise from top left:
YUS rainforest plant. *Photo by Lisa Dabek, TKCP.*
Long claws help tree kangaroos to climb. *Photo by Lisa Dabek, TKCP.*
Native orchids in YUS. *Photo by Danny Samandingke, TKCP.*
Coffee cherries. *Photo by Dono Ogate, TKCP.*

Greg Parrott

The Reeve Family

Patti Savoy

Adam and Catherine Schaeffer

Gena Shurtleff

Gary Smith and Kathleen Kemper

Lisa Tiedt

Utah's Hogle Zoo

Lauren Wyckoff

ZooParc de Beauval

_____

Up to $999

Anonymous (7)

Richard Abel and Roberta Berner

Hannah Ahmed

Harriet Allen

Avery and Marcia Aten

Robert Bailey

Dominique Bideau

Richard Biribauer

Barbara Birney

The Boeing Company Matching Gifts

Victor Bozzo

John and Sarah Brooks

Mylene Brooks

Barbara Christensen and Jeff Meyer

Leonard and Sharon Clemeson

Susan Cohen

Stephan Coonrod and Cheryl Clark

Gabriel Cronin

Kim Daly-Crews

Sophie Danforth

Brian Darley

James DeBonis

Daniel Dechert

Patrick Dessalle

Scott Dew and Colleen Hanlon

Tamara DiCaprio

Laurie Ann and C. Bert Dudley

George and Barbara Ermentrout

Donna and Steve Estes Antebi

Charles and Rose Ann Finkel

Janice and William Fischel

Harmony Frazier and Michael Breen

Deena Fuller

James Galbraith

Mary Gillmore

Madeleine Hagen

Edie and Brian Hall

Susan Hall

George and Carol Harell

Ryan and Heather Hawk

Nancy and Paul Hawkes

Sheila and Earl Horowitz

Rochelle Howe and Jonathan Greene

Mike Kaputa and Suzanne Tomassi

Ken Katsumoto

Jenny Kim and Stephen Sun

Jeanne and Jason Kinnard

Amy Kitchener

Yoko Kobayashi

Nicole Labrecque

BJ and Nayna Laird

Jacob Langley

Monica Lieb

Lincoln Children's Zoo

David and Lois Madsen

Lindsay Malone

Chris McFarlane and Arianne Foulks

Christine McKnight

Gary Mozel

Val and Laird Muraoka

Judy Nyman-Schaaf

Oaklawn Farm Zoo

Darrin OBrien

Anne Palaszewski

Christopher Pepin and Ken Miller

Craig Pepin

Mimi Polk Gitlin

Jeremy Potash

Helen Ralph

Helen Ramirez

Roberta Roberts

Rohrbach Family

Kimberly Sanders

Santa Fe College Foundation, Inc.

Carol and Seymour Sarnoff

Benjamin Schweinhart

Ellen Sciutto

Patricia and Scott Sebelsky

Judie and Rick Steenberg

Anne Stein

Laurie Stewart

Jonathan and Tiffany Sweet

TCS & Starquest Expeditions

Steven Thornton and Nancy Ostrander

UBS Financial Services Matching Gifts

Russ White

Kevin and Jo Wilhelm

Mike and Jan Williams

Ann P. Wyckoff

Christy Wyckoff

Stacie and Joseph Zane

Jacob Zimmerman

Deceased

Thanks

The Tree Kangaroo Conservation Program would like to give thanks to all of our partners, colleagues, supporters, and friends who have contributed to the success of the Tree Kangaroo Conservation Program. Your collaboration and support make a tremendous difference for the people and wildlife of YUS.

TKCP is the signature conservation program at Woodland Park Zoo, and relies on the contributions of many departments and colleagues within the zoo. We give special thanks to the following departments and colleagues at WPZ:

Woodland Park Zoo Board of Directors

Woodland Park Zoo President and CEO Dr. Deborah Jensen

Woodland Park Zoo Field Conservation Department including Dr. Fred Koontz, Bobbi Miller and Dr. Robert Long

Woodland Park Zoo volunteers Courtney Baxter, Judy Nyman-Schaaf and Trish Watson

Woodland Park Zoo Finance Department including Valerie Krueger, Celeste Sabers, Nathan Ricard, Marilyn Spring and Carol Baroff

Woodland Park Zoo Creative Services team including Kelly Hampson and Misty Fried

Woodland Park Zoo Development Department including Paul Balle, Lorna Chin, Barbara Folger, Paris Jones, Anne Knapp, Kate Neville, Susan Okazaki and Sarah Valentine

Woodland Park Zoo Human Resources Department

Woodland Park Zoo Communications Department including Rebecca Whitham, Ryan Hawk, Caileigh Robertson, Laura Lockard and Gigi Allianic

Woodland Park Zoo Animal Management Department including Deanna Ramirez, Beth Carlyle-Askew, Wendy Gardner, Dr. Jennifer Pramuk, Helen Shewman and Dr. Nancy Hawkes

Woodland Park Zoo Animal Health Department

Woodland Park Zoo Education Department including Jamie Creola, Eli Weiss, Carla Bitter, Katie Remine, Jenny Mears, Scott Vance, Kathryn Owen, Kim Haas and WPZ docents

Woodland Park Zoo Horticulture Department including David Selk

Woodland Park Zoo Retail Programs including staff members Terry Blumer, Ashley Hoover, Lisa Bagnell and all other staff that have supported our work

Woodland Park Zoo Security Department including Trinidad Alcaraz

 Singsing celebration in Isan village. *Photo by Namo Yaoro, TKCP.*
Matschie's tree kangaroos. *Photo by Dr. Rob Liddell.*

TKCP is also grateful for the support of our partners and friends in Papua New Guinea and across the globe. We extend a special thanks to:

PNG Department of Environment and Conservation with special thanks to Minister of Environment the Honorable John Pundari, Deputy Secretary Kay Kalim, John Michael, Arthur Ganubella, Benside Thomas, James Sabi and Madeline Lahari

Honorable Kasiga Kelly Naru, Governor of Morobe Province

Honorable Member for Kabwum District, Mr. Bob Dadae and Kabwum District Administrator Mr. David Kitenge

Morobe Provincial Government including Mr. Taikone Gwakoro, Mr. Robin Kiki, Mr. Keith Jiram, Mr. Micah Yawing

Yus Local Level Government including Yus LLG President Honorable Epemu Kiwenu, the Councillors, Magistrates, and Manager Mr. Fidel Yapenare

Wasu Local Level Government including Wasu LLG President, Mr. Petrus Yasing, the Councillors and Magistrates

Simon Simboki, Extension Officer with Wasu LLG

YUS Conservation Organization

Isan Community for hosting the YUS CO General meeting, CAMC, TKCP Staff meeting, and the Land-use Planning launch celebration

Saburong Community for hosting the TKCP Staff meeting and Ranger meeting

YUS Teachers, Headmasters and School Board

YUS Community Health Workers and Village Birth Attendants

PNG Forest Research Institute and Professor Simon Saulei, Mr. Wake Yelu, and Mr. Anton Lata

Dr. Larry Orsak and Dr. Kulala Mulung from the Department of Forestry, and Joy Sahumlal from the Department of Distance Learning at the University of Technology of Lae

Balob Teachers College for their ongoing partnership in our YUS teacher scholarship program, in particular Mr. Jerry Hendingao and Mr. Lenggkepe Zomgoreng

Morobe Agricultural Show Committee

Georgia Kaipu of the National Research Institute (NRI)

PNG Institute for Biological Research (IBR)

Gwen Sissiou of PNG's Office of Climate Change and Development

The Research and Conservation Foundation (RCF) of PNG

Nik Sekhran, Joseph D'Cruz, Johan Robinson, Julianne Zeidler, Tamalis Akus, Christie Mahap and Gwen Maru of UNDP

The Nature Conservancy PNG especially Francis Hurahura and Clement Kipa

Adelbert Cocoa Cooperative Society

Karl Aglai, John Kabuba, and Daniel Kilagi of PNG's Coffee Industry Corporation

Monpi Coffee Exports Ltd. including Adam Kline

Monpi Cocoa Sustainable Services, especially Hannah Wheaton

Queen Emma Chocolates, especially David Peate and Willie Wong

The farmers of Marawasa village for hosting the agro-forestry nursery exchange visit

Steven Boting and the Etep Rural Hospital health team

Sister Baleb Wahazoka, Former Family Health Services Coordinator at Morobe Provincial Health Department

Lae International Hotel

Dr. Jane Mogina, PNG LNG project, ExxonMobil

North Coast Aviation, Missionary Aviation Fellowship, and Summer Institute of Linguistics for air transport

The Leahy family

U.S. Ambassador to Papua New Guinea Hon. Walter North and the U.S. Embassy

Maurice Knight and Pankaja Panda at the USAID Mission in Papua New Guinea

Atlas of Living Australia (ALA) Biodiversity Volunteers, especially Paul Flemons

Dr. Jared Diamond for guidance and for contributing to the YUS Conservation Endowment fundraising efforts

Simon Passingan of Barefoot Community Services

Simon Rollinson of Pacific Island Projects

Marleen Knijff

Atherton GIS (ATGIS) and Alistair Hart

Philp Schouteten at Baste Design

ECOM Agroindustrial Asia Pte Ltd including Stephen Bannister

Linzon Zamang

The BALANCED Project collaborators, including Dr. Joan Castro (PATH Philippines), Linda Bruce and Janet Edmond (CI)

Conservation International including CI-Global Conservation Fund, Russell Mittermeier, Jennifer Morris, Chris Stone, Angela Kirkman, Martine Culbertson, Valeria Martinez, Olivier Langrand and David Mitchell

Bernard Maladina

The Centre for Environmental Law and Community Rights (CELCOR)

James Cook University including Dr. Andrew Krockenberger and Gabriel Porolak

Dr. Sasha Aikhenvaldand and Hannah Sarvasy

Margit Cianelli

Tree Kangaroo and Mammal Group of the Atherton Tablelands, Australia

The United Nations Development Programme and the Equator Initiative

Daniel Shewmaker and Caffe Vita

Nathan Palmer-Royston and Theo Chocolate

Carl Darnell of Chinook Medical Gear for generous support of our field equipment needs

Jamie Bechtel and New Course

Foundations of Success

Jacque Blessington and the AZA Tree Kangaroo Species Survival Plan

AZA Marsupial and Monotreme Taxon Advisory Group

Dr. Britt Yamamoto and iLEAP, Seattle, Washington, USA

Harriet Allen, Washington Department of Fish and Wildlife

Doctors Rob and Marti Liddell, Blair Brooks and Nancy Philips, and Travis Austin

Steve Whisker, ExxonMobil

Mike Kaputa and Suzanne Tomassi

Susan Barkan, University of Washington

David Gillison

Carol and Seymour Sarnoff

Robert Plotnick and Gay Jensen as champions for our program in Seattle

TKCP-PNG Board of Directors Dr. Lisa Dabek (WPZ), Dr. Bruce Beehler (Smithsonian Institution); Mr. Francis Hurahura (The Nature Conservancy PNG); Mr. Zachary Wells (CI); and Ms. Mikal Nolan (TKCP-PNG)

Henrietta Philips, TKCP archivist

Henri Dabek, who passed away in 2014

Samandingke Daum Duenane, who passed away in 2014

Mr. Herry Ereyu, in memory of one of TKCP's most committed field carriers

Most importantly, we would like to thank the people of YUS for their unending dedication to being stewards of their environment, and for their gracious hospitality while we are in their villages.

Huon Peninsula, Papua New Guinea

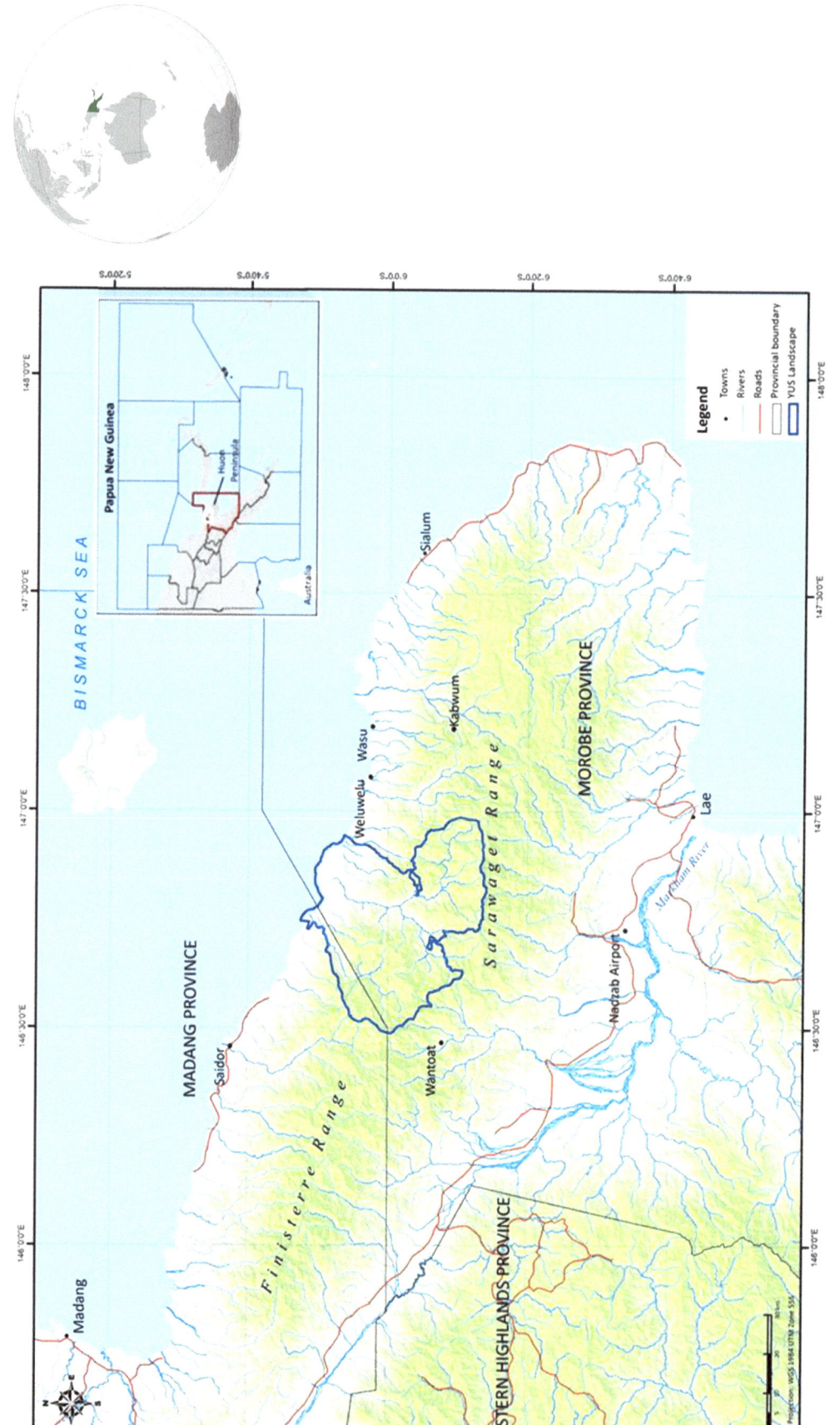

Map by Karau Kuna Jr, Tree Kangaroo Conservation Program

YUS Conservation Area, Papua New Guinea

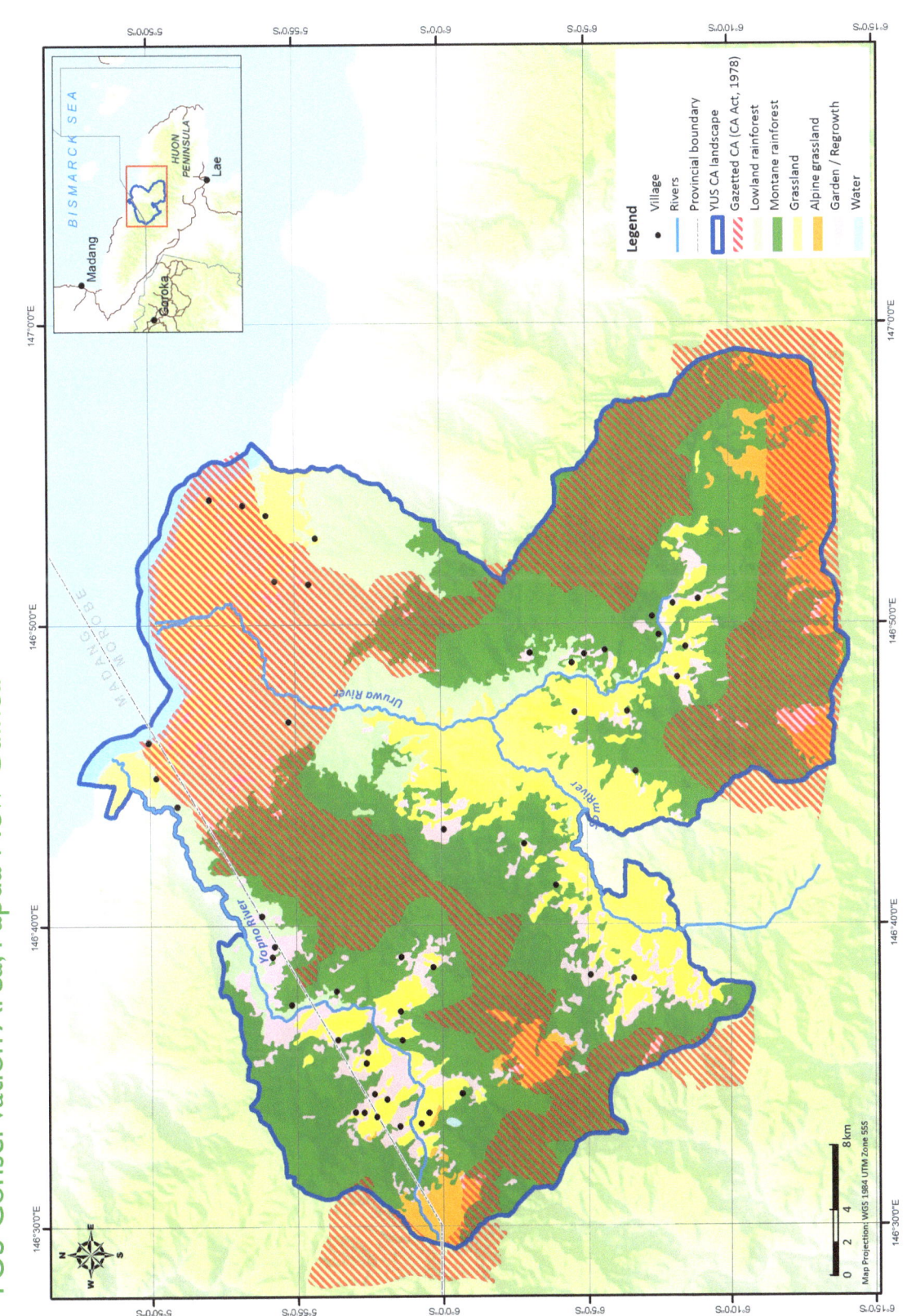

Legend

- • Village
- Rivers
- Provincial boundary
- YUS CA landscape
- Gazetted CA (CA Act, 1978)
- Lowland rainforest
- Montane rainforest
- Grassland
- Alpine grassland
- Garden / Regrowth
- Water

BISMARCK SEA

HUON PENINSULA

Madang

Lae

Goroka

MADANG / MOROBE

Uruwa River

Yopno River

Som River

Map Projection: WGS 1984 UTM Zone 55S

0 2 4 8 km

Map by Karau Kuna Jr, Tree Kangaroo Conservation Program